*Animal
Communication
by Pheromones*

Animal Communication by Pheromones

H. H. Shorey

Department of Entomology
University of California
Riverside, California

ACADEMIC PRESS   New York   San Francisco   London   1976

A Subsidiary of Harcourt Brace Jovanovich, Publishers

ACADEMIC PRESS, INC.
111 Fifth Avenue, New York, New York 10003

United Kingdom Edition published by
ACADEMIC PRESS, INC. (LONDON) LTD.
24/28 Oval Road, London NW1

Library of Congress Cataloging in Publication Data

Shorey, Harry H
        Animal communication by pheromones.

        Bibliography: p.
        Includes index.
        1.   Animal communication.   2.   Pheromones.   I.   Title.
QL776.S54   591.5'9   75-44765
ISBN 0−12−640450−X

Contents

Preface   vii

1 *Introduction*
   Text   1

2 *The Pheromonal Communication System*
   2.1 Emission of Pheromone   7
   2.2 Transport of Pheromone   14
   2.3 Reception of Pheromone   16

3 *Mechanisms of Movement and*
   Orientation to Pheromone Sources
   3.1 Orientation with Respect to a Pheromone Gradient   19
   3.2 Terrestrial Odor-Trail Following   20
   3.3 Aerial Odor-Trail Following   22
   3.4 Distances of Pheromone Communication   27

4 *Recognition Pheromone Behavior*
   4.1 Recognition of Individuals   37
   4.2 Recognition of Status   38
   4.3 Recognition of Group   40
   4.4 Recognition of Home   41
   4.5 Recognition of Home Range   43

5 *Aggregation Pheromone Behavior*
   5.1 Exploitation of Sources of Food   45
   5.2 Aggregation prior to Sexual Behavior   50
   5.3 Aggregation prior to Aggressive Behavior   52
   5.4 Maintenance of Societal and Family Groups   52
   5.5 Colonization of Habitats   55
   5.6 Other Aggregation Behavior   60

6 *Dispersion Pheromone Behavior*
    6.1   Maintenance of Optimal Interindividual Spacing   65
    6.2   Maintenance of Territories   67
    6.3   Dispersion in Response to Alarm Pheromones   68
    6.4   Antiaggregation   71

7 *Aggression Pheromone Behavior*
    7.1   Stimulation of Aggression toward an Individual of
          Another Species   75
    7.2   Stimulation of Aggression toward a Conspecific   80
    7.3   Inhibition of Aggression toward a Conspecific   82
    7.4   Release of Pheromone as an Aggressive Act   83

8 *Sex Pheromone Behavior*
    8.1   Stimulation of Aggregation   85
    8.2   Stimulation of Courtship and Copulation   93
    8.3   Hierarchies of Sex Pheromone Behavior   98
    8.4   Human Sex Pheromones   101

9 *Environmental and Physiological*
Control of Sex Pheromone Behavior
    9.1   Environmental Control   105
    9.2   Physiological Control   108

10 *Sex Pheromones and Reproductive Isolation*
    Text   113

11 *Evolution of Pheromonal Communication*
    Text   117

    Bibliography   123
    Taxonomic Index   159
    Subject Index   163

Preface

Pheromones are chemicals, either odors or taste substances, that are released by organisms into the environment, where they serve as messages
to others of the same species. Although humans exude a great variety of
chemicals, they make little conscious use of this potential means for communication with one another. On the other hand, pheromones are widely
used within much of the rest of the animal kingdom in a great variety of
species, ranging from primitive protozoans to higher primates, as a
primary means for transmitting information. Depending on the particular species involved and the situation in which it finds itself at the time,
pheromones may be used for attracting a mating partner or for stimulating that partner to copulate, for directing others to suitable food or
resting sites, for causing others to stay away when staying away is appropriate, or for a variety of other behavioral functions.

The scientific literature dealing with pheromones has expanded enor-
mously during recent years as have the number of reviews which have
proliferated in symposium volumes and in collections of chapters prepared by individual contributors. Although most of the review articles
have dealt with insect pheromones, the importance of mammalian pheromones has received increased recognition in recent years, and a number
of reviews concerning this group have also been published. However,
with the exception of a book by Martin Jacobson entitled "Insect Sex
Pheromones" (published by Academic Press in 1972), no single-authored
monograph concerning pheromones has appeared.

I have felt for some time that the information concerning pheromone
communication within the entire animal kingdom should be reviewed,
digested, and presented in a cohesive manner. This book represents my
attempt to perform this task. It is mainly directed toward an assessment
of how the behavior of animals is controlled and influenced by pheromone communication. Attention to individual taxa, such as worms, in-

sects, or fish, is minimized. Instead, an attempt has been made to generalize the diverse behaviors exhibited by animals when they are engaged in pheromone communication and to group together discussions of both primitive and advanced animals when they are using pheromone communication in a similar manner for such behavioral functions as sex, aggression, feeding, and recognition of other individuals. I have also attempted to draw attention to some of the interesting and specific pheromone behaviors that have evolved in particular animal species in relation to their particular ways of life.

Placing relatively simple invertebrates and complex vertebrates in the same generalized scheme involves the risk of making all these greatly diverse types of animals seem too much alike. However, despite this possibility, I felt this type of scheme of presentation valuable in achieving a mainly behavioral view of pheromone communication in the animal kingdom.

I wish to acknowledge a number of my colleagues who assisted me during the preparation of this book. J. S. Gaston prepared Figures 6, 8, 9, 10, 17, 18, and 26, and L. B. Bjostad prepared the schematic drawings of chemical molecules. P. A. Murray assisted in library research and in cataloging the literature. A. E. Colwell, J. F. Bollinger, and L. K. Gaston offered valuable advice concerning the substance of the manuscript. H. R. Bowman did all of the typing and the laborious collating of material. Finally, my children, Tom, Russell, Diane, and Hal, provided patience and encouragement during the years before and during preparation of this work; this book is dedicated to them.

H. H. Shorey

1

Introduction

Most animals must communicate with others of their own species. The necessity for communication is obvious for animals, such as honeybees, rabbits, and man, that live in complex societies. Communication is the mechanism through which these social animals interact with each other and by which they are organized according to their relative statuses and functions.

Although the need is less obvious, communication is also essential in the lives of most animals that appear to live alone. However, for these solitary animals this communication often takes place only at certain critical times during their lives. For example, prior to mating in bisexual species, communication must be used to bring the two sexes—or at least their gametes—together. Once the two individuals have come together, additional communication is usually necessary to stimulate and guide the process of copulation or the union of gametes. A premating communication system may operate over a long range and be obvious to the human investigator, or it may operate only at very close range, perhaps following chance contact of the individuals, and be less obvious. Even in the latter situation, it appears unlikely that the entire sequence leading to mating occurs in any species without the occurrence of some communication between the animals involved.

As will be indicated throughout this book, many types of communication may be of vital importance in the lives of "solitary" animals, including the exchange of messages that keep them from coming together at inappropriate times. Thus, if a definition of sociality includes the communication of information among individuals of a given species, then almost all animals are social on a continuum from the "solitary" species to those that live in complex societies.

Considerable disagreement occurs concerning a precise definition of the term "communication" (115). I will adhere to a broad definition patterned after Wilson (710): Biological communication entails the release of one or more stimuli by one individual that alters the likelihood of reaction by another, with the reaction being of benefit to the stimulus emitter, the stimulus receiver, or both. A stimulus may act directly, eliciting an overt behavioral reaction in the receiving individual, or, it may only cause a change in the receiving animal's threshold of responsiveness to other stimuli.

A variety of stimulus modalities and associated sense organs may be used by communicating animals, the major ones being usually classified as chemical (olfactory or gustatory), mechanical (tactile or sonic), and radiational (light perception or visual). These major communication channels occur in widely diverse groups of animals, ranging from protozoans through the complex higher vertebrates. The chemical sense is very primitive (253, 710); it probably allowed primordial single-celled organisms to locate food in their environments and, as sexuality developed, to locate each other for exchange of genetic material. As discussed later (Chapter 11), the development of chemical communication among individual free-living cells was probably a necessary precursor to the evolution of metazoan animals, which consist of interacting groups of cells.

Chemical communication among animals of the same species is widespread and occurs throughout the animal kingdom. In fact, in many diverse groups of animals (with some notable exceptions, including the birds and higher primates), chemical communication appears to be the major channel for exchange of information. However, one must avoid overgeneralizing this point; various emphases on the other modalities of communication occur also. Since each species has evolved its mode of communication according to its individualistic way of life, channels for transfer of information between individuals of a given species are varied and could involve any one of the modalities of communication or even a mixture of these modalities.

This book is restricted to a consideration of chemical communication between individuals of the same species. The term "ectohormone," coined by Bethe (57) to describe these chemicals, is contradictory (329), with "ecto" meaning "external" and "hormone" referring to an "internal secretion." The chemicals are also often called "attractants" or, depending on their biological function, "sex attractants," "feeding attractants," etc. Such terms are too general in that they do not denote that the chemicals are used for communication between individuals of a given species. Also, although attraction of one animal to another may be the most obvious (to a human observer) result of chemical communication, a chemical that stimulates attraction may cause additional responses, such as copulatory behavior, once the animals are together.

In 1959, Karlson, Lüscher, and Butenandt (329, 330) proposed the term "pheromone" to represent a chemical(s) used for communication between individuals of a given species. The term is derived from the Greek "pherein" meaning "to transfer" and "hormon" meaning "to excite." Although "pheromone" was criticized early on etymological grounds, it has received wide acceptance and is now used throughout the world in the scientific literature (63). With incorporation of some minor modifications as proposed by Kalmus (324), a pheromone can be defined as a chemical or a mixture of chemicals that is released to the exterior by an organism and that causes one or more specific reactions in a receiving organism of the same species. Thus, if two or more chemicals are released together and cause a specific reaction, the chemical mixture must be regarded as a pheromone.

A more rigorous definition might require that a pheromone be synthesized *de novo* in specialized glands in the producing animal. To adhere to this rigor is difficult and may be unnecessarily restrictive. A continuum probably exists, from chemicals that are produced *de novo* and secreted from specialized glands, to chemicals that are metabolic byproducts of related substances that have been ingested, to chemicals that are metabolized within the animal to their biologically active state by symbionts such as intestinal bacteria, to chemicals that are ingested and released later in their original form. In most cases, the chemical identity of a pheromone and the biosynthetic mechanisms that led to the production of the pheromone are unknown. In the practical world, an investigator who observes that one or more unknown chemicals released by an animal cause a behavioral response in another animal of the same species calls the chemical message a pheromone.

Pheromones have been divided by Wilson (706) into two broad categories. A "releaser pheromone" triggers an almost immediate behavioral response in the receiving animal, with the response probably being mediated entirely by nervous system pathways. An overt response to the pheromone sometimes might not be seen. The response might entail only a change in the threshold of reactivity of the animal to some other environmental stimulus, such as the sight or sound of a potential mating partner. Thus, using a neurophysiological approach (at least for higher animals), we might more accurately say that a releaser pheromone stimulates specific chemosensory organs to relay action-potential-coded messages via their sensory neurons to the central nervous system. There the messages modify (increase or reduce) the likelihood that messages will leave the central nervous system via specific motor neurons, causing the animal to respond in a correspondingly specific way.

"Primer pheromones," constituting Wilson's second category, are probably also detected by chemosensory organs that relay appropriate messages to the central nervous system. However, the response is not a direct behavioral reaction. Instead, it entails a relatively enduring reorganization of the physiology of the receiving animal. Most known primer pheromones cause changes, presumably mediated by hormonal systems, in the developmental or reproductive processes of the receiving individuals. Examples of primer pheromones are male-produced chemicals that modify the reproductive state of female mice (105) and queen-produced chemicals that inhibit ovary development in worker honeybees (216). Primer pheromones will not be dealt with beyond this brief description, the focus of this book being on those relatively instantaneous changes in behavioral response (or responsiveness) stimulated by releaser pheromones.

Man's research interests tend to be anthropocentric, and most early investigations into animal communication were involved with those sensory modalities that are maximized in the human species, vision and audition. However, man has now become more aware that most of the rest of the animal kingdom appears to maximize chemical communication. A recent redirection of research interest has resulted in an explosive increase in our appreciation and knowledge concerning animal pheromones.

A practical factor has stimulated the entry of many investigators into the pheromone field and—of great importance—has provided an infusion of funds from agencies that sponsor research. Especially among the insects, certain pest species appear to be highly dependent on pheromone communication for their survival. Such pheromone communication systems include those that bring the two sexes together prior to mating and those that direct species mates to appropriate food or breeding sites (Chapter 5). After acquiring an intimate knowledge of the normal pheromone communication behavior in pest species, man can attempt to manipulate the behavior to his own advantage. He can modify a previously adaptive behavior, such as the orientation of an animal to a pheromone-emitting mate, into a nonadaptive one, such as entry into a trap baited with the same pheromone. Pheromones have been utilized for some time as bait in traps used for survey of the distribution and abundance of pest insects. To date, no practical pheromone-based system has been proved successful as a direct means for obtaining effective control of pests. However, considerable research is presently underway in this area and important breakthroughs enabling pheromonal control of insect behavior, and thus of pest insect populations, seem imminent (65).

Microanalytical chemistry is an area of research that has added considerable impetus to the pheromone field and has made possible the practical applications mentioned above. The first chemical identification of a pheromone used in the premating behavior of moths was that of the female silkworm. Identification of this pheromone was completed in 1959 (117) following many years of effort. The identification was based on about 12 mg of pheromone isolated from the abdominal glands of 500,000 female moths. The second identification, reported in 1966 (48) for the female cabbage looper moth, was based on about 5 mg of pheromone collected from 2500 female moths. Since that time, identifications of female moth sex pheromones have been reported at a rapidly increasing rate, totalling over 40 by 1976. With recent advances in instrumental techniques for characterizing small quantities of chemical, many identifications are now accomplished on a few hundred micrograms of pheromone.

The study of pheromones cuts across and utilizes the tools of many diverse disciplines, i.e., morphology, physiology, ecology, behavior, chemistry, and pest control. No encyclopedic attempt has been made in this text to include reference to every relevant research article. In those cases in which a large number of original articles document similar phenomena in related animal groups, only certain of the references are mentioned. Some aspects of pheromone communication are covered very briefly or not at all. The morphology of glands that produce pheromones and of sensory organs that perceive pheromones is mentioned ony briefly in Chapter 2. Chemical techniques are not discussed. However, the chemical structures of many animal pheromones are presented in the various chapters that deal with the use of these chemicals in animal communication. The use of pheromones in pest control is not mentioned at all, although considerable research effort is occurring in this area. Interested readers who desire additional information regarding these aspects are referred to a number of recent reviews (65, 66, 301, 309, 473, 479, 591, 602, 654).

This book is organized along behavioral lines, illustrating the various types of behavior stimulated by pheromones, the mechanisms by which the behaviors come about, and the adaptive advantages which accrue from the behaviors. Most of the behaviors may be separated into two general categories. First, a pheromone may cause locomotion and/or orientation (steering) responses that bring an animal toward or away from the pheromone source. These responses have the effect of causing animals to aggregate (Chapter 5) or disperse (Chapter 6). Second, a pheromone may stimulate or inhibit other responses such as sexual behavior (Chapter 8) or aggression (Chapter 7)—usually when the animal is close to the source of pheromone emission. Often, the same pheromone that causes animals to aggregate near its source also stimulates the additional close-range behavior. Likewise, a pheromone that causes animals to disperse may also inhibit other behaviors, such as aggression. Those pheromones that cause multiple behavioral reactions (i.e., aggregation followed by sexual behavior or aggression) are discussed more than once, appearing in each of the appropriate chapters. This organizational scheme is somewhat different from that employed by other authors who have categorized pheromones along more functional lines, such as those resulting in alarm behavior (all locomotory and close-range behaviors exhibited in response to pheromones released during times of danger) and sexual behavior (all locomotory and close-range behaviors exhibited in response to pheromones released by potential mating partners) (119, 706, 710).

2

The Pheromonal Communication System

A typical communication system consists of three components: a mechanism for emitting the message, a medium through which the message is transmitted from the point of initiation to that of reception, and a mechanism for receiving the message. In the case of pheromone communication, the emitting component is often a glandular organ associated with specialized devices that transfer the chemical molecules into the surrounding medium. The reception component is an olfactory (smell) or gustatory (taste) sensory organ. The medium may be air or water, depending largely on the way of life of the species involved and the medium in which it lives. However, in the case of those pheromones that are liberated onto the surface of one animal and that are perceived by gustation by a second animal that makes direct contact with the first, the medium becomes essentially nonexistent.

2.1 Emission of Pheromone

2.1.1 EMISSION FROM ANIMALS

Although the locations and morphological features of the pheromone-producing glands of most animal species remain unknown, these organs have been extensively studied in insects and mammals. In fact, the description by morphologists of the glandular structures found on a number of animal species has preceded the demonstration by behaviorists that the structures do indeed secrete pheromones.

The types and locations of pheromone glands found in insects and mammals are as varied as are the many communication functions attributable to the pheromones. Depending on the species, the glands may be found on the head, thorax, abdomen, or legs (Figs. 1, 2). In insects, the glands may also be found on the wings. Often, a single animal has a number of glands in a variety of locations.

Most pheromone glands of insects and mammals are composed of groups of modified epidermal cells. A gland may consist of a single layer

7

FIG. 1. Mature male marmoset monkeys, *Callithrix j. jacchus,* display their external genitalia, which are covered with scent glands. This behavior is a signal of aggressive threat and may provide visual as well as olfactory information to nearby males. (Courtesy of G. Epple.)

of cells on the exterior of the animal, or the gland may be complex and associated with internal reservoirs. Elaborate secondary devices associated with the gland may have the function of disseminating the pheromone into the surrounding medium. For instance, the pheromone glands on the tarsi of the black-tailed deer are associated with erectile tufts of hairs

FIG. 2. Female cabbage looper moths cling to vertical surfaces or to the lower portion of horizontal surfaces when exposing the sex pheromone gland located at the end of their abdomens. (Courtesy of S. Gothilf.)

which greatly increase the surface area for evaporation of the pheromone molecules into the air (440). Similarly, male moths and butterflies frequently possess modified scales resembling hairs. The scales are enclosed in pouches which contain the pheromone glands. The pouches evaginate, erecting the pheromone-coated scales into the air at the time of premating communication between males and females (Fig. 3) (64). The ability to expose or erect pheromone-disseminating devices or to eject pheromone from reservoirs enables an animal to control the time of its pheromone emission.

2.1.2 EMISSION FROM OBJECTS

In many cases, the pheromone does not volatilize directly from the emitting animal. Instead, the emitter deposits the secretion on objects or on the substratum, establishing a scent mark. The deposited pheromone serves as the site for evaporation of the chemical into the air, and responses of other animals of the same species may be directed toward the marked area. Thus, scent marks have the advantage that the communicative process can occur even in the absence of the pheromone emitter,

FIG. 3. A pinned museum specimen of the moth, *Creatonetus gangis,* showing the coremata that are inflated with air and extended from the terminal portion of the male abdomen during courtship and which may serve as devices for dissemination of pheromone. (Courtesy of M. C. Birch.)

which may have deposited the secretion and then moved on to other activities.

Scent-marking behavior has reached its peak in the social insects and the mammals. Both groups have developed a great diversity of specialized glands and methods for depositing the pheromone. Sometimes a single species has several different glandular sources used in depositing scent marks. The function of most insect scent-marking behavior seems to be well understood and will be considered in later chapters.

In contrast, although scent-marking is performed by most mammalian species, the behavioral functions of the marking are generally not known. Little controlled experimentation has been conducted, and the functions are mainly inferred from observational evidence (496, 704). Johnson (309) has indicated that mammalian scent marks might act as

    1. A deterrent or a substitute for aggression to warn conspecifics away from occupied territory

    2. A sex attractant or stimulant

    3. A system for labeling the habitat for an animal's own use in orientation or to maintain a sense of familiarity with an area

    4. An indicator of individual identity, perhaps including information regarding such factors as sexual status, age, and dominance

    5. An alarm signal to conspecifics

    6. An indicator of population size

A list of similar functions has been proposed by Mykytowyz (450). In some mammalian species, more than one of the above functions might be served by a single scent mark. No attempt will be made to interpret the listing further at this point. Those cases where experimental evidence supports the proposed functions will be discussed in later chapters.

Pheromones used in mammalian scent-marking may be deposited with the urine or feces or by the direct contact of skin (cutaneous) glands with the substratum (93, 309, 352, 450, 724). The pheromonal property of urine or feces is probably not attributable to the odor of these substances per se. Instead, cutaneous glands are often associated with the channels for urination or defecation and the urine or feces may be thus scent-marked as they pass from the body.

Pheromones may be deposited by either active or passive marking. Passive marking occurs through the nonspecific liberation of a pheromone and its application to the substratum as the animal travels through its environment. For example, the pedal glands found between the bases of the hooves of ruminants apparently mark the environment with their secretion as the animals move about (93).

Active marking occurs at a time and in a place that is selected by the animal. The marking takes place through direct gland-to-object contact

FIG. 4. A male marmoset monkey, *Leontopithecus rosalia rosalia,* scent-marks his perch with the secretion from his circumgenital glands. (Courtesy of G. Epple.)

or through the selective application of pheromone to the urine or feces when they are discharged. Some examples of active marking follow (93, 219, 268, 450). Mongooses and marmoset monkeys smear branches with secretion from their anal glands and circumgenital glands, respectively (Fig. 4). Ungulates apply pheromone to trees by directly rubbing their preorbital or retrocornal glands on the trunk or branches. Mountain goats spread the product of their supraoccipital glands on tufts of grass. Beavers rub their pheromone glands on hillocks of mud or objects projecting above the water. Rabbits use their chin glands to mark a variety of localized features in their environment such as logs, branches (Fig. 5), blades of grass, the entrances to their burrow, their own or others' fecal pellets, their females, and their young.

FIG. 5. Rabbits use their chin glands to scent-mark objects, such as branches, in their territories. (Courtesy of E. R. Hesterman and R. Mykytowycz.)

Marking with urine by some animal species may involve elaborate behavior patterns. The Canidae (dogs, wolves, etc.) establish common urination points to which visits are made by all local inhabitants; however, only the males participate in social urination and contribute to the odor source (93, 268, 352, 724). Both brown bears and bison urinate on the ground. They then roll in the urine, following which they rub their bodies against trees, depositing scent marks (Fig. 6).

Anal glands, apparently for marking the feces with pheromone, have been identified in over 100 species of mammals (450). Many mammals accumulate piles of feces at certain spots, where they probably serve a communicative function. The sizes of these communal dunghills vary among the different species. Those of the hyena may cover an area of a quarter of an acre (407).

Recently, behaviorally active components of the pheromone produced in scent-marking glands have been identified for two mammalian species. Phenylacetic acid (1) is found in the ventral gland of male Mongolian gerbils (658). Both male and female gerbils possess ventral glands, with which they mark a variety of objects in their environment. Marking in the male has been proposed to have a role in exploratory, territorial, and social dominance behavior. A similar chemical, isovaleric acid (2), was

1   **2**

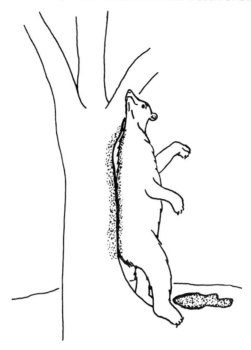

FIG. 6. Brown bears often deposit urine on the ground and then roll in it. They then rub against trees, presumably depositing scent marks. [Reprinted by permission from Hediger, H. (1949). Säugiter Territonen und ihre Markierung. *Bijdr. Dierk.* **28,** 172–184.]

identified as one of the components produced in the subauricular gland of the male pronghorn antelope (443). The gland is rubbed directly on vegetation. Among the proposed functions of the deposited scent marks are identification of individuals, territorial behavior, threat behavior during aggressive encounters between bucks, and sexual behavior. Because of the typical complexity of mammalian pheromones, it seems likely that many more chemical components will eventually be identified from the scent-marking glands of these species (Chapter 4).

2.2 Transport of Pheromone

The medium for transmission of pheromone molecules from the emitting to the receiving animal is essentially always air or water. Even for animals that live in the soil, the molecules probably are conveyed by some combination of the water film that surrounds the soil particles and the air spaces between them. The two major methods for chemical dispersion within air or water are diffusion, attributable to the Brownian movement of molecules, and passive transport, attributable to the movement or flow of the medium.

Diffusion is of practical importance in distribution of odor molecules only when the medium is not flowing. In still air, only about 10% of molecules of molecular weight 100 to 300 (encompassing most airborne pheromones) travel 1 cm from their source in 1 second. Diffusion rates in water average 10^4 to 10^5 slower than those in air. However, the size range of molecules that might function as waterborne pheromones is much greater than that possible for airborne pheromones, ranging from a molecular weight of 17 for the hydroxyl ion which has been implicated as a sex pheromone of a nematode species (632) to over 10,000 for the polypeptides or proteins which may be used as pheromones by a number of aquatic animal species (710). Regardless of molecular size, fewer than 0.001% of the waterborne molecules would be expected to diffuse more than 1 cm from their source in 1 second.

This consideration of diffusion as a mechanism for dispersion of the pheromone message through the medium must be related to two other parameters: the rate of release of pheromone from the emitting animal and the molecular concentration that is needed for stimulation of the receiving animal (Chapter 3, Section 3.4). For example, a pheromone-emitting female cabbage looper moth releases approximately 10^{11} molecules into the air in 1 second (619). Although only a small proportion of the total, many millions of molecules will diffuse far from their source during the next second (Table 1), and the molecular concentration may still be sufficiently high to stimulate an olfactorily sensitive male moth located several cm away. The role of diffusion in disseminating phero-mone messages is considered further in Chapter 3, Section 3.4.

Air and water are rarely static. They flow because of prevailing winds or stream or tidal movement. Even in the absence of prevailing flows, the medium is almost always moving. Unevenly heated surfaces give rise to convection currents which create local movement of the medium. In addition, except at very low velocities, the medium rarely flows in a smooth, laminar fashion (640). Moving air and water are characterized

Table 1. INFLUENCE OF DIFFUSION IN DISSEMINATING PHEROMONE MOLE-CULES AWAY FROM A SOURCE[a]

Time following pheromone release (seconds)	Number of molecules/mm³ of air at indicated distances from pheromone source (cm)			
	1	2	3	10
1	6×10^6	3×10^3	< 1	< 1
3	6×10^6	5×10^5	8×10^3	< 1
10	3×10^6	8×10^5	2×10^5	< 1

[a] Source instantaneously releases 10^{11} molecules of molecular weight 200 into still air.

by turbulence, giving rise to lateral displacement of any pheromone molecules that might be carried. When the medium is in motion, then, the prevailing direction of flow causes the major mass transport of pheromone molecules, turbulence causes a lateral spread of molecules from their predicted downwind or downstream location, and diffusion is of negligible importance in causing further molecular dispersion.

2.3 Reception of Pheromone

In most of the pheromone communication systems studied to date, the chemicals are perceived by means of olfaction rather than gustation. The olfactory receptors of both vertebrates and invertebrates are remarkably similar in fundamental structure, even though their superficial appearances are very different (Fig. 7) (569, 570, 628, 629). Those of vertebrates are typically found in the nasal cavity; the dendrites of the olfactory sensory neurons extend into the lumen of the cavity and are bathed in a mucus film. Invertebrate olfactory receptors may be found on various external parts of the body. The olfactory receptors of arthropods are usually located on paired antennae on the head, although they may also be found in other areas, especially associated with the mouthparts and the genitalia. The dendrites of the sensory cells of arthropod olfactory receptors often extend into specialized sensilla or sensory hairs. The surface of each hair contains numerous pores that lead to a liquid in the hair, bathing the dendrites. Thus, in both vertebrates and invertebrates

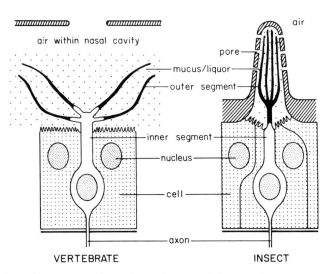

air within nasal cavity

air

pore

mucus/liquor

outer segment

inner segment

nucleus

cell

axon

VERTEBRATE

INSECT

FIG. 7. Schematic representations of vertebrate and insect olfactory receptors, with possibly analogous structures drawn similarly. [Adapted from Steinbrecht (628).]

16

the odor molecules apparently must be absorbed in a liquid bathing the dendrites and move to the acceptor surfaces on the dendrites by diffusion.

Humans are most familiar with olfaction involving air as the transmitting medium. However, water can also serve as the medium. For example, fish smell chemicals dissolved in the water that they circulate through their nasal cavity. The fish nasal cavity is completely analogous to that of terrestrial vertebrates and is distinct from the oral cavity (17).

Some animals, such as dogs and male moths, are extremely sensitive to olfactory stimuli. A single pheromone molecule contacting the receptive surface of an olfactory sense cell of the male silkworm moth is sufficient to cause the cell to initiate a nerve impulse in its axon that leads to the olfactory center in the brain (322, 571, 629). In such olfactorily sensitive species, the sensory cells can function as "molecule counting" devices.

3

Mechanisms of Movement and Orientation to Pheromone Sources

The behavioral mechanisms that cause animals to approach or to remain near a pheromone source are poorly understood. However, some information is available, and an attempt will be made here to formulate some generalities as to how pheromones may cause aggregation behavior. Most of the critical research has been conducted on terrestrial insects, primarily ants, beetles, and moths, and the following discussion is based mainly on these groups. It is likely that many of the proposed mechanisms pertain also to a great variety of invertebrate and vertebrate species that live in either terrestrial or aquatic environments.

Two general classes of reactions are involved in aggregation near a pheromone source. One class is related to the speed of locomotion and is called the *orthokinesis* reaction (202). Pheromone stimulation may at first cause the animal to move and may control the rapidity of the locomotion. As the animal nears or arrives at a favorable odor source, the rapidity may become inversely proportional to the pheromone concentration, resulting in arrestment or cessation of locomotion. The second class relates to the orientation or steering of the body axis of the stimulated animal and is called the *taxis* reaction (202). When the taxis reaction occurs at the same time as the orthokinesis reaction, the animal may move toward the pheromone source. As will be discussed below, the steering may be directed with respect to a gradient of the pheromone itself, or into the direction of flow of air or water, or toward visual objects perceived in the environment.

3.1 Orientation with Respect to a Pheromone Gradient

The steep concentration gradient of odor molecules that is sometimes found close to a pheromone source may allow the animal to orient chemotactically, by sensing the direction of the gradient and steering its body accordingly. When the sensing of the gradient is accomplished by the animal's simultaneous comparison of the differing odor concentrations at two points in space, the subsequent steering is called *chemotropotaxis*. Such a simultaneous comparison is often made possible by the bilateral symmetry of the olfactory organs on the receiving animal—the paired antennae of an arthropod or nostrils of a vertebrate. If one of the paired organs receives higher stimulation than the other, the responding animal turns toward that side. As inferred earlier, the turbulence associated with moving air rapidly breaks up odor gradients, and chemotropotactic orientation of animals to such gradients is therefore probably impossible at distances greater than a few centimeters from the odor source (Chapter 2, Section 2.2). Factors that control the potential distance of chemotropotactic orientation at any one moment are the integrity and steepness of the gradient, and the distance between the olfactory receptor organs on the responding animal. A large animal having widely spaced receptors might be expected to detect a gradient when further from the odor source than would a smaller one.

A second method by which an animal might detect a pheromone gradient is that of successive scanning across the odor field with the whole body or part of the body, such as head or antennae. If the animal then orients its body toward the area in which the greatest odor concentration was sensed, the steering is known as *chemoklinotaxis*. Even with a badly disrupted gradient, one can visualize how this mechanism might allow orientation to an odor source over a distance as great as a few meters, as when a deer turns its head from one side to the other while sniffing the air.

Some animals might use either chemotactic mechanism, depending on the circumstances. Bees orienting to a nearby odor source may sense the chemical gradient chemotropotactically, by comparing the signals received from the two antennae, or chemoklinotactically, by moving one or both antennae across the odor field (388, 391).

3.2 Terrestrial Odor-Trail Following

Some animals deposit pheromone on the substratum in the form of a continuous line or discontinuous streaks. The pheromone evaporates into the air or dissolves into water, forming an elongate odor trail. These trails are called "terrestrial" to emphasize the fact that they are deposited on a solid substratum, which may be on dry land or under water, and are typically followed by individuals that are maintaining contact with the same substratum while they move. In still air or water, the active space, defined as that space in which the odor molecules are at a concentration above the behavioral threshold of an animal responding to the pheromone (712), takes the form of a semicylinder. The axis of the semicylinder lies along the line of chemical on the substratum, and a concentration gradient of molecules extends laterally from the axis in any cross section of the trail (Fig. 8).

Terrestrial odor-trail forming and following behavior occurs in such diverse animals as ants, snails, snakes, and dogs. The trail-following individual typically does not need to sense the deposited pheromone by gustation, but senses the odor above the trail by olfaction. It seems likely that a trail-following animal maintains contact with the trail and keeps its body near the central axis of the trail by detecting and steering with respect to the lateral concentration gradient (194, 254, 255, 705, 712). Perception by the animal that it is losing odor contact as it diverges off to the right or left of the axis of the trail probably leads to chemotactic steering reactions that cause it to return toward or across the axis (Fig. 9). If the animal is engaged in forward locomotion at the same time that it performs the lateral corrective steering, its progression along the trail should assume a sinusoidal or zigzag character. The animal, therefore, may maintain contact with the trail by continuously sensing when it is losing contact with it. The zigzag following pattern is seen in a variety of terrestrial trail-following species including dogs and ants. A further

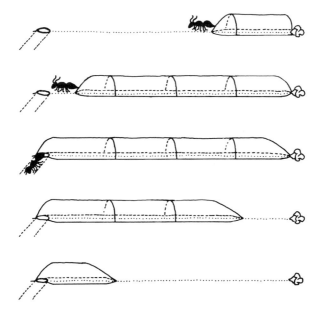

FIG. 8. Diagrammatic representation of the active space of pheromone in the air above a terrestrial trail deposited by an imported fire ant worker while returning from a food source to its nest. The successive diagrams show the formation and fade-out of the trail with the passage of time. [Adapted from Wilson and Bossert (712).]

FIG. 9. Sinusoidal movement of an ant following a line of terrestrial trail pheromone drawn on a horizontal surface. [Adapted from Hangartner (254).]

benefit derived from the zigzag motion relates to a lessening of sensory adaptation, which would probably occur rapidly if the animal were exposed to the relatively constant level of odorant above the midline of the trail (712).

The terrestrial pheromone trails of some animal species are inherently polarized, providing information as to which direction along the trail the follower should proceed. The trails laid by certain aquatic pulmonate snails when they move toward the water surface to replenish their air supply are polarized; other snails coming upon a trail proceed in the same direction as the individual that deposited the trail (700). The factor that indicates the polarity of the trail is not known. It may not even be olfactory; the follower might detect directionality through gustatory contact with the trail surface. The terrestrial trails used by ants to designate the route to sources of food are not polarized. However, ants and other animal species following nonpolarized trails may still proceed in the correct direction by sensing familiar features in the environment or by using photomenotaxis, the sun-compass steering reaction that causes them to proceed at a predetermined angle with relation to the position of the sun (282).

3.3 Aerial Odor-Trail Following

Although the following discussion relates mainly to aerial systems, the same principles probably also pertain to aquatic systems. When moving air sweeps past a source of pheromone, an elongate cloud or plume of odor molecules is produced on the downwind side. The pheromone plume is not uniform and linear; instead, it is disrupted and filamentous in nature, consisting of pulses of high and low pheromone densities (Fig. 10). The aerial pheromone plume has been called a trail by Butler (119). This terminology appears appropriate, because many animals have behavioral mechanisms that enable them to follow an odor trail through three-dimensional space. Our present knowledge concerning the trail-following mechanisms is based on diverse and often conflicting experimental results and observations. However, enough information is available to permit some speculation and generalization concerning two probable mechanisms involving anemotaxis (or rheotaxis) and chemotaxis.

FIG. 10. The filamentous nature of an odor plume that is formed by the continuous evaporation of a pheromone from a point source into air moving in the direction of the arrow.

3.3.1 ANEMOTACTIC AND RHEOTACTIC MECHANISMS

Most investigators agree that an animal could not possibly sense which direction to move toward an odor source located many meters or kilometers away by detecting a concentration gradient of molecules radiating out from the source. The gradient would be too shallow at those great distances, and it would be totally disrupted by atmospheric turbulence and flow. However, the animal might be stimulated by the odor to orient its body into the wind or current and thus approach the odor source (158, 194, 257, 335, 336, 579, 582, 592). This type of orientation, made with respect to the direction of medium flow, is known as anemotaxis (wind-steering) for animals orienting in the air and rheotaxis (current-steering) for those orienting in water.

An animal having physical contact with a solid substratum might be able to detect the direction of wind or current flow by sensing the direction to which the mechanoreceptors on its body are deflected. This method of detection will not work for an airborne or free-swimming animal. The mechanoreceptors of flying or swimming animals can only indicate the speed or direction that the animals are moving with respect to the medium; they do not indicate the speed or direction that the medium is moving with respect to the substratum. However, a flying or swimming animal can sense the direction of movement of the medium if it maintains visual contact with the substratum. If an animal that is supposedly flying straight ahead sees that it is sideslipping to the left, relative to the ground beneath it, then the wind must be blowing from its right side. Corrective steering reactions toward the right might then be made, until the substratum appears to pass from front to rear, directly under the animal's body, causing it to be oriented upwind. Visually based anemotaxis has been rigorously demonstrated for males of two moth species approaching sources of female sex pheromone (336). Similar systems have been demonstrated also for the upwind orientation of flying *Drosophila melanogaster* and mosquitoes in response to nonpheromone odors (334, 335, 722).

Considering the disrupted nature of the aerial odor trail, a flying animal can be expected to lose contact with the trail from time to time. The animal then might cease its upwind orientation and make back and forth crosswind movements in a manner that would maximize the likelihood that it would reenter the trail and be stimulated to steer anemotactically again. The direction of the crosswind movements would also, presumably, be visually based. They have been demonstrated experimentally for male moths of several species, at the time that the moths lost contact with an aerial trail of female sex pheromone (336, 665). If a male of the Mediterranean flour moth does not reenter the pheromone trail after several crosswind casts, it drops back downwind and then resumes its crosswind flight pattern (665).

3.3.2 CHEMOTACTIC MECHANISMS

The movement of insects flying along aerial trails of pheromone often consists of a sequence of sinusoidal or zigzag oscillations in a lateral plane (54, 111, 143, 172, 192, 238, 272, 277, 341, 394, 547). This zigzag progression of the flying insects is remarkably similar to that discussed earlier for animals that run along terrestrial odor trails. This observation leads to the hypothesis that the mechanisms of odor-trail following in these two groups might have an important element in common. Perhaps both terrestrial and aerial animals sense when they are diverging laterally from the highest average pheromone concentration found near the central axis of the trail and are entering the lower concentrations near the boundaries of the active space, whereupon they are caused to turn back toward the higher concentration. Such a chemotactic mechanism might continuously reorient the animal toward the trail's central axis. However, if the animal were far downwind from the site of pheromone emission and if the trail were quite wide, it is difficult to visualize how classical chemotropotaxis—instantaneous comparison of the chemical concentration sensed by paired receptor organs—could enable this lateral orientation. But, it is likely that some form of chemoklinotaxis—successive comparison of concentrations in different points of space over time—could operate. A medium-sized moth can fly several meters per second and might, depending on the sophistication of its central nervous system, be able to integrate considerable information concerning changes in odor concentration over time.

Instead of sensing changes in molecular density as it moves through the aerial odor trail, the animal might sense changes in the frequency at which it encounters pulses of pheromone. As discussed earlier, the aerial trail is highly filamentous. Both the average density of odor molecules and the average frequency of filaments decrease as an animal moves laterally from the central axis as well as downwind from the source. Wright (721) proposed that the direction of flight of an animal orienting to an odor source is guided by the frequency with which it contacts the pulses of high density of odor molecules within the aerial trail. If the pulses are contacted at an increasing frequency, indicating that the animal is proceeding toward the odor or toward the central axis, no turns are made. If the pulses are encountered at decreasing frequency, the animal tends to turn until it again senses an increasing frequency. This idea, although it appears plausible and should be studied further, was later abandoned by Wright (723) because of negative experimental results.

Some limited experimental evidence supports the hypothesis that a chemotactic mechanism (or modified chemotactic mechanism based on pulses) is utilized for aerial odor-trail following. Farkas and Shorey (192) demonstrated that male pink bollworm moths proceed in their usual zig-zag manner along an aerial trail of female sex pheromone even when the trail is momentarily suspended in still air. However, their experiments did not demonstrate a polarization of the trail, and it is possible that other cues (perhaps visual) enabled the males to fly in the correct direction, toward the pheromone source (193, 243, 336). In another experiment, Shorey and Farkas (594) removed the wings from male cabbage looper moths, which normally fly within an aerial trail toward pheromone-emitting females. A line of the pheromone was drawn on a horizontal substratum, in a manner similar to that used in experiments intended to demonstrate terrestrial odor-trail following in ants. The wingless males were released on the substratum and followed the terrestrial odor trail for a considerable distance on foot.

It seems reasonable that both of the proposed mechanisms for aerial odor-trail following may often operate simultaneously, with chemotaxis maintaining the animal within the lateral active space of the trail and anemotaxis providing polarity and orienting the animal toward the pheromone source. Other mechanisms, not yet proposed, are likely to be involved also. Animal orientation to distant odor sources is undoubtedly based on complex behavioral mechanisms that may operate separately or together, depending on the species and on a number of physiological and environmental variables.

3.3.3 MECHANISMS USED WHEN NEAR THE PHEROMONE SOURCE

As the animal approaches the pheromone source, it encounters a progressively narrower aerial odor trail and a higher average molecular concentration. These factors may lead to a shift from those behavioral mechanisms discussed above for approach to a distant odor source to mechanisms appropriate for approach to a nearby source. The latter mechanisms may include visual orientation to nearby surfaces or to the image of the pheromone-emitting animal itself, short-range anemotactic or chemotactic orientation, and arrestment of locomotion.

3.3.3.1 Visual Guidance Visual orientation to nearby objects has been observed in insects of a number of species when they are stimulated by high pheromone concentrations. Many bark beetles, upon encountering high concentrations of their aggregation pheromone (Chapter 5, Section 5.5) tend to visually orient to vertical objects. The behavior seems to be adaptive, since a vertical object in their normal environment is likely to be the trunk of the tree from which the pheromone trail originates (212, 213, 685, 689). Likewise, males of the pink bollworm moth, when exposed to a high concentration of female sex pheromone, tend to hover adjacent to vertical objects, presumably maximizing the likelihood that they will find pheromone-releasing females located on plants (195).

At very short range (probably within a few centimeters of a pheromone-emitting female), the high odorous concentration may stimulate a male moth to visually orient directly to the female's image (120, 162, 171, 277, 285, 598, 665). The pheromone is often essential to this behavior, and the males of some species do not appear to observe a nearby female at all if they are not stimulated by the appropriate odor.

3.3.3.2 Short-Range Anemotaxis and Chemotaxis Chemotactic gradient-following mechanisms may also operate within a few centimeters of the pheromone source to steer the animal in the right direction. Among flying insects, wing vibration by the responding individual probably aids in the gradient detection, being equivalent to sniffing by mammals. Many insects, when exposed to a high pheromone concentration, turn their bodies from side to side while in hovering flight or while vibrating their wings and clinging to the substratum. The vibrating wings cause the local air mass in front of the insect to be drawn across its antennae and exhausted to the rear. Schneider (568) has speculated that the turning of the insects may be a very advantageous behavior, allowing them to consecutively sample the pheromone concentrations from various directions in space.

Even the direction of the copulatory attempt of a male insect may be oriented chemotactically. Shorey (589) removed one antenna from male cabbage looper moths and then presented them with filter paper impregnated with female sex pheromone. The males everted their genitalia and attempted to copulate with the treated paper. However, the direction toward which their genitalia were thrust was biased toward the side of the body having an intact antenna.

Females of some species of moths frequently vibrate their wings when clinging to a surface and releasing sex pheromone (592). This behavior results in local pheromone-bearing air currents behind the females. The air velocity may be considerable; it has been measured as 15 cm/second, 1 m behind pheromone-releasing female cabbage looper moths (318). The local air stream may enable males to use anemotactic mechanisms in their short-range approach to the pheromone source. Also, female cabbage looper moths can sense the prevailing wind velocity and have a greater tendency to vibrate their wings, producing the local air currents, as the prevailing velocity decreases from 4 to 0 m/second (318).

3.3.3.3 Arrestment of Locomotion A behavioral mechanism must be present to slow the speed of the moving animal as it approaches a pheromone source and to cause it to cease locomotion altogether when it arrives at the source. This mechanism—at least in certain flying insects—apparently involves a decrease in the rate of forward locomotion in proportion to the increasing concentration of pheromone perceived (47, 90, 194, 195, 665). Although the pheromone at low concentration may act as an initial stimulus to cause the animal to start moving along the odor trail, at high concentrations it acts as a stimulus to inhibit locomotion. When a male pink bollworm moth approaches a pheromone-emitting female, the slowing of locomotion is accomplished in part by a reduction in the moth's wing-beat frequency and in part by an increase in the angles through which it turns during the zigzag approach flight (195).

When the animal arrives at the pheromone source and stops moving, the high odorous concentration may also act as an arrestant, inhibiting it from starting to move again (44, 88, 213, 353, 546, 689, 716, 718, 719).

3.4 Distances of Pheromone Communication

The potential distance over which pheromone communication between two animals can occur is determined by a number of interacting factors. These include the rate at which pheromone is released by the emitter, the lowest concentration of pheromone molecules that is detectable by the responder, and properties of the medium that cause the molecules to dilute to the minimal detectable concentration at some distance from the pheromone source. It is apparent that the communication distance will increase with either an increase in molecular release rate or an increase in sensitivity of the responder.

Bossert and Wilson (92) provide calculations and examples that relate to potential communication distances in four different situations: (1) pheromone emitted as a discrete pulse into still air, (2) pheromone emitted continuously into still air, (3) pheromone emitted from a terrestrial trail into still air, and (4) pheromone emitted from a point source into moving air, forming an aerial trail.

3.4.1 PHEROMONE EMITTED AS A DISCRETE PULSE INTO STILL AIR

Ant alarm pheromones are used in communication systems that often operate over very small distances and for short periods of time. The pheromones are released by ants that are injured or that perceive a threatening situation. In the still air in or near an ant nest, the pheromone may spread mainly by diffusion and may stimulate other ants within the radius of the active space to perform behaviors appropriate to the situation (Chapters 5–7). If all the alarm pheromone contained in a worker ant of the species *Pogonomyrmex badius* were liberated instantaneously through the crushing of its head, the zone of the active space containing a molecular concentration above the threshold necessary for perception by other workers would attain a maximum radius of about 6 cm in 13 seconds (712). Within 35 seconds the active space, and thus its effect, would fade out completely, because of continued diffusion. If the worker were merely alarmed and not crushed, the amount of pheromone released would undoubtedly be lower and the active space would be correspondingly smaller and fade away more rapidly. Thus, with a small disturbance in the nest, the number of workers influenced by the pheromone message is relatively small, the disturbance is rapidly subdued, and the signal rapidly dissipates. However, with a larger disturbance, a correspondingly larger amount of pheromone released causes the activation of a larger number of workers over a longer period of time. The intensity, duration, and area covered by the communication system, then, appears to be highly adaptive. The rapid fade-out time of the signal prevents a lingering effect, which might disrupt colony activities after the threatening situation is removed or which might affect a larger part of the colony than is necessary.

3.4.2 PHEROMONE EMITTED CONTINUOUSLY INTO STILL AIR

Continuous emission of pheromone into still air would, theoretically, lead to an increase in the active space up to some maximum radius which depends on the pheromone release rate and the threshold concentration detectable by a responding animal. Potential communication distances for this type of situation have received little study. Usually, the distances must be restricted to within a few centimeters of the pheromone source, because odor concentration gradients would normally not remain intact over greater distances (Chapter 3, Section 3.1).

3.4.3 PHEROMONE EMITTED FROM A TERRESTRIAL TRAIL INTO STILL AIR

The maximum possible length, the lateral dimensions, and the fade-out time of the active space of a terrestrial odor trail are interrelated and depend on the amount and volatility of the pheromone deposited on the substratum and the olfactory threshold of the responding animal. These factors have become adjusted according to the selective pressures affecting any given species so that they are appropriate to the way of life of that species. Some trails are highly persistent and may be very long. For example, Texas leaf-cutting ants harvest leaves and bring them back to their nests, where they use them as a substrate for their fungus gardens. Their terrestrial trails may lead to permanent sources of vegetation 100 or more m from the nest, and these trails may be followed for many months (Fig. 11) (274, 433). At the other extreme, the terrestrial trails of the imported fire ant are very ephemeral (Fig. 9). This species forages for insects and other small bits of food in close proximity to its nest. The trails have such a rapid fade-out time that their effective distance is less than 1 m (705, 706). The short fade-out time eliminates "noise" or confusion from the system, with the trails being self-eliminating after a source of nearby food has been removed.

3.4.4 AERIAL PHEROMONE TRAILS

Pheromones are in many cases emitted continuously from point sources into moving air. The pheromone source may be a scent gland located on the releasing animal or it may be a scent mark left by an animal on some substratum in the environment.

Dilution of a chemical in moving air is largely a result of the velocity and turbulence of the air mass. In general, as the air velocity decreases, the potential communication distance increases. This occurs because the number of molecules released into any given volume of air is inversely proportional to the velocity at which the air sweeps past the pheromone source.

Based on equations derived by Sutton (640), Bossert and Wilson (92) determined that the mean active space containing an above-threshold concentration of molecules assumes the shape of a semiellipsoid extending downwind from a pheromone source located near ground level (Fig. 12). The reader should be aware that the Bossert and Wilson model is based on a long-time average dispersion pattern of molecules downwind

FIG. 11. Trails of the Texas leaf-cutting ant may extend 100 m or more from the nest, be 20 cm or more wide, and last 6 months or longer, depending on the persistence of the foraging source. (Courtesy of John C. Moser.)

from an odor source. At any one instant in the real world the phero-mone trail does not look like the calculated semiellipsoid, but instead looks like the sinuous, filamentous trail shown in Fig. 10. The active space in the instantaneous trail may at times extend well beyond the end of the semiellipsoid, because the calculated pattern is the average of all the instantaneous trails. This averaging, however, allows for approximate calculations of communication distances. The following formula may be

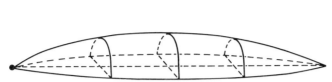

FIG. 12. Calculated mean active space of pheromone in the air downwind from a continuously releasing source located at ground level. [Adapted from Wilson and Bossert (712).]

used to determine a theoretical mean maximum distance of pheromone communication in moving air (92):

$$X = (8Q/vK)^{4/7}$$

where X = communication distance (cm), Q = pheromone release rate (μg/second), K = pheromone response threshold (μg/cm^3 of air), and v = wind velocity (cm/second).

Using the formula, Bossert and Wilson (92) calculated that the theoretical mean maximum communication distance between a pheromone-releasing female and a responding male of the gypsy moth is 4.6 km when the wind velocity is 1 m/second. The actual communication distance, however, is probably much less than this, because the calculations may have been based on a great underestimation of K, the male response threshold (144).

Recently, Sower et al. (619) calculated that the mean pheromone emission rate of female cabbage looper moths is 7 ng/minute (range <1 to >20 ng/minute) and the mean threshold dilution of pheromone needed to cause 50% of the males to respond is 80 molecules per cubic millimeter of air (range <8 to >800 molecules per cubic millimeter of air). Using the Bossert and Wilson formula, they then calculated that the mean maximum communication distances for this species are 5, 20, and 80 m at air velocities of 5, 0.5, and 0.05 m per second, respectively. In reality, communication distances among animals of any given species must be highly variable. Considering the variations in female release rates, male thresholds, and air velocities, potential communication distances for individual pairs of the cabbage looper moth might accurately be described as ranging from less than 1 to somewhat over 100 m.

Successful pheromone communication requires not only that pheromone molecules be carried by moving air to the vicinity of the responder, but also that the responder move to the vicinity of the emitter while uninterrupted pheromone release takes place. Although certain species release pheromone more-or-less continuously, others may have only brief periods of release. Females of the cabbage looper moth appear to sense the prevailing wind velocity and adjust the durations of their phero-mone-release periods accordingly (318). The length of continuous emis-sion is decreased from about 20 minutes at very low velocities (<0.1 m/second) to 5 minutes at a velocity of 3 m/second. This varied duration of release seems to be adaptive in that it increases the potential distance of pheromone communication at the low velocities, when the pheromone trail moves most slowly downwind from the female. The 3 m/second velocity represents the maximum velocity at which successful communi-cation is possible; the male moth flight speed is also approximately 3 m/second, so progress toward the pheromone source at greater veloc-ities would be impossible. Thus, there are practical constraints on the successful communication distance, caused by interactions between the duration of pheromone release, male flight speed, and wind velocity. These constraints were used by Sower *et al.* (622) in deriving the follow-ing formula for estimating the maximum possible distance of successful communication:

$$X = vt \ (r-v)/r$$

where X = communication distance (cm), t = time that the pheromone is emitted without interruption (seconds), r = approach speed of the re-sponder (cm/second), and v = wind velocity (cm/second). Concurrent application of this formula and the Bossert and Wilson formula allows

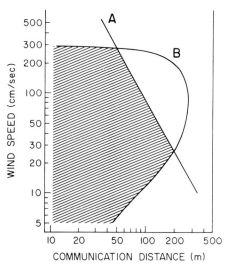

FIG. 13. Maximum theoretical distances of sex pheromone communication between female and male cabbage looper moths, estimated by (A) the formula of Bossert and Wilson, and (B) the formula of Sower *et al.* [Adapted from Sower *et al.* (622).]

estimation of the maximum theoretical distance of successful pheromone communication (Fig. 13).

Data needed for these calculations, including the rate and duration of pheromone release, the response threshold and locomotion rate of the responder, and wind velocity, are rarely available. Therefore, most estimates of communication distances are based on direct observations in the field. Known numbers of animals are marked in a distinctive way, released into the environment at known distances from a pheromone source, and later recaptured at the source. Using this method, a number of investigators have estimated that some moth species have potential communication distances of greater than 1 (and sometimes greater than 10) km. However, a serious weakness is inherent in this mark–recapture method in that there is no way to determine how far an animal might have moved in the absence of odorous stimulation before it finally perceived an above-threshold molecular concentration and then followed the trail the remainder of the way to the source (231, 592).

Before orientation mechanisms such as anemotaxis or chemotactic trail following were proposed, investigators of moth behavior—including the great French naturalist Fabre (652)—were unable to conceive how an odorous system could enable approach of males to a distant, sexually responsive female. Therefore, from time to time, other mechanisms of distance communication have been proposed, especially systems based on detection by the males of radiant energy emitted by the females. However, to date there has been no demonstration that any message other than the pheromone molecules emitted by the female moths is effective over considerable distances (167).

4

Recognition Pheromone Behavior

Although pheromones often act as stimuli that cause the perceiving animal to exhibit overt behavioral reactions, they may in some cases play a more subtle role, conveying information about the identity or about some characteristic of the emitting animal. These chemicals have been termed "identification," "recognition," or "appraisal" pheromones. Although the terms have various shades of meaning, they will be used here interchangeably. In general, the term recognition pheromone appears to be most apt, signifying that the receiving animal perceives the pheromone (and thus the pheromone emitter) as something that was previously known. These pheromones may be used to recognize: (1) individual animals of the same species, (2) the physiological or social status of species mates, (3) assemblages of individuals, constituting colonies or social groups, (4) the animal's own home or nest, or (5) the animal's home range.

A problem often arises in attempting to differentiate whether a pheromone truly is used for the recognition of one of the above categories, or whether it only acts as a stimulus, inducing the perceiving animal to perform a characteristic response when it perceives the appropriate odor. Can an ant that follows the terrestrial odor trail (Chapter 5, Section 5.1) made by other members of its species, but does not follow that of a related species, be said to recognize or identify its species' pheromone? Probably it cannot. The ant might not even perceive the pheromone of the related species, or, if it does detect the pheromone, the chemical composition of the odor trail of the wrong species might not constitute the correct stimulus to induce trail-following behavior. Similarly, a ram that is stimulated to engage in precopulatory behavior upon perceiving a sex pheromone (Chapter 8, Section 8.2) emitted by an estrus ewe might be said to use pheromones to identify the sex of its species mates. However, the same ram might not be sexually aroused in the presence of other rams, merely because it perceives no appropriate stimulating odor. Also, a rat pup approaches a lactating female that is emitting a "maternal" pheromone (Chapter 5, Section 5.4), whereas it does not approach a nonlactating female that is not releasing the pheromone. Each of the above examples indicates possible inappropriate uses of the terms recognition or identification, which imply that the animal passively discriminates between other individuals or characteristics of other individuals on the basis of their pheromones. Especially in the vertebrates, a continuum probably exists between passive recognition and classical stimulus–response behavior, and in most cases investigators have not distinguished between these two categories.

Despite the above discussion of the difficulties that arise in attempting to define "recognition," there is no question that vertebrates often discriminate among individuals that they have previously known by means of the odors characteristic of those individuals. Therefore, Wilson (710) proposed that pheromones of vertebrates tend to be composed of complex mixtures of chemicals, and that those of invertebrates mainly consist of single chemicals or relatively simple mixtures. He speculated that the differences relate to the fact that the social behavior of most vertebrate species is "personal," whereas that of invertebrates is characteristically "impersonal." Vertebrate social behavior is proposed to be largely based on the recognition of individuals and of prior relationships with those individuals, such as parent, subordinate, or mate. Thus, each animal in those vertebrate species that use pheromones for individual recognition might be expected to have a characteristic odor comprised of many chemical components whose proportions can be varied. On the other hand, much invertebrate social behavior is accomplished by stimulus–response reactions leading to mating, dispersion, aggregation, etc., and for these reactions one or a few key chemicals might suffice. This contrast by Wilson seems reasonable, although until more information becomes available it should be accepted with reservations. As Wilson (710) points out himself, a notable invertebrate exception may be the colony odors of social insects; these odors are specific to colonies and are often variable through time within given colonies. Also, although recognition of individuals by a vertebrate probably requires a large medley of chemical ingredients in a pheromone, the supposed recognition of certain other vertebrate characteristics such as sex, parental relationship, or social status might require only one or a few key chemicals that stimulate the appropriate behaviors.

The beaver contains at least 50 chemicals in its castoreum gland (342), lending credence to the multiplicity of chemicals in a vertebrate pheromone. Also, the secretions of the anal, chin, and inguinal glands which are used in the scent-marking behavior of the rabbit, *Oryctolagus cuniculus*, vary in composition from individual to individual as well as between the sexes (228). On the other hand, Bergström and Löfqvist (50, 51) recently demonstrated that the secretion produced in the Dufour's gland of the ant, *Camponotus ligniperda,* contains at least 41 chemicals. As more sophisticated chemical investigations are made, the supposedly simple sex or aggregation pheromones of many insects are found to be composed of complex chemical mixtures (72; Chapters 5 and 8).

4.1 Recognition of Individuals

Most literature on the role of pheromones in enabling individual recognition pertains to mammals, although there is some information indicating that pheromones may be of importance in this regard in many fish and certain invertebrate species.

Olfactory recognition often appears to be important in establishing the bond between a mammalian mother and her offspring (105, 442). The mother may ignore or even attack young animals of the same species that do not possess the odor characteristic of her own young. In fact, some rodent females, after being experimentally deprived of their sense of smell, have been observed to kill and eat their own pups. The young may also distinguish their own mother from other females by recognizing her distinctive odor. An early imprinting with the correct odor may influence not only the young animal's future recognition of and relationships with its mother, but also its selection of a mate having a similar odor when it becomes adult.

A mammalian pheromone used for individual recognition may volatilize directly from the body of the animal, or it may be deposited onto a substratum as a scent mark. In those species that release the odor from their bodies, a common behavior upon the meeting of two individuals is the direct approach to, and the sniffing of, a characteristic glandular area (186, 187, 189, 310, 440, 566). For example, black-tailed deer sniff the tarsal scent gland of other deer (440). The major chemical component found in the tarsal gland secretion of male deer has been identified as *cis*-4-hydroxydec-6-enoic acid lactone (3) (113); other components must be present also in order for the secretion to permit the differentiation of individuals.

3

A scent mark has the advantage of allowing an animal to identify the previous presence of either a known or an unknown individual of the same species in a particular area. Also, a mark may in some cases be used to self-identify, i.e., to recognize that the sniffing animal had been at that location at some earlier time. This self-identification is likely a major factor involved in an animal's recognition of its nest or home range (Chapter 4, Sections 4.4 and 4.5).

A common behavioral response of animals when near a scent mark is an approach to the marked object and the deposition of a new scent mark on it. This behavior is commonly observed in the urination patterns of the domestic male dog. In some highly communal species, such as the rabbit, the same marking points are visited regularly and remarked by all members of a social group. These scent marks may act as loci for the general exchange of information such as the individual identity, as well as the age, sex, breeding condition, and social status of the marking animals (309, 440, 441).

Few studies have been conducted to determine how many different individuals can be specifically identified on the basis of scent alone. In mice, the number is at least 18 (323).

The use of pheromones for individual identification among fish has been studied extensively in the yellow bullhead (17, 663). The bullheads can recognize other individuals of their own species by means of odor alone. After a pair of fish has been allowed to interact, a small amount of water conditioned by one of them is sufficient to elicit the appropriate response from the other. For example, if one fish had established dominance over the other, the subordinate flees when it perceives the dominant's pheromone.

Another fish, the blind goby, *Typhlogobius californiensis*, lives in burrows as single pairs of a male and female (396). A goby within its burrow recognizes invading gobies of the same sex by means of a pheromone, and the male will fight to the death with an invading male or the female with an invading female. This example does not indicate an identification of various individuals, but rather a differentiation of the odor of all other gobies of the same sex from that of the resident male or female. A similar situation is seen in an invertebrate, the desert wood louse, *Hemilepistus reaumuri* (393). A male and female live together in a burrow. Either of the pair can enter the burrow without being attacked by the other resident. However, all other members of the species are recognized by means of a pheromone and are driven away.

4.2 Recognition of Status

Vertebrates are frequently characterized as being able to recognize the physiological status—including age, sex, stage of estrus cycle, and copulatory readiness—of species mates by means of pheromones. As mentioned earlier, most observations do not discriminate as to whether true recognition of the physiological status is involved, or simply a direct stimulatory effect of the pheromone on the recipient. The response to—and, perhaps, the recognition of—pheromones released from animals of differing physiological states will be considered in the following chapters.

Social status among vertebrates usually refers to the relative position of a particular animal in a dominance hierarchy of coexisting individuals. The degree of dominance or submissiveness of an animal may be recognized by others by means of pheromones (309, 450). We might speculate that the levels of dominance are communicated by qualitative or quantitative differences in the blends of chemicals constituting the pheromone released by dominant as opposed to submissive animals. Some morphological and behavioral evidence indicates that quantitative differences are most likely. The more dominant individuals in mammalian social groups are characterized as having larger scent-marking glands, higher pheromone secretory rates, and a higher frequency of scent-marking than subordinate individuals (Fig. 14) (106, 276, 440, 441, 451, 656, 657).

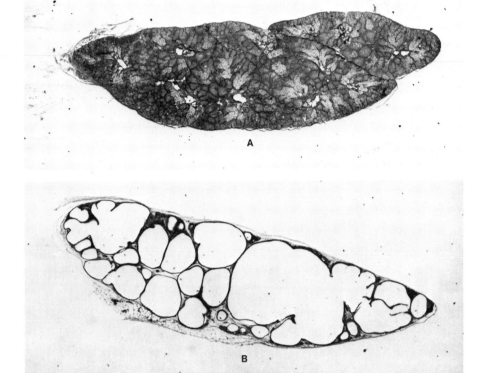

Fig. 14. Preputial gland of a dominant (A) and subordinate (B) male mouse. (Courtesy of N. W. Nowell.)

Changes in social status are often associated with hormonal changes within the animal, and the hormone titer may directly affect glandular size and secretory rate. Dominant male animals maintain higher androgen levels than subordinates. Castration of the dominant animals causes a dimunition of pheromone gland activity; injection of the castrates with androgen causes gland growth and resumption of secretion (178, 449).

4.3 Recognition of Group

Mammals of many species live together in groups of interacting individuals. The members of a group recognize and tolerate each other, but they are generally intolerant of members of foreign groups. The recognition of group membership is often accomplished by one or more pheromones. Typically, an animal having an odor that is foreign to a particular group is attacked if it enters the territory occupied by that group, whereas an animal having a pheromone known to the group is accepted (Chapter 7). This pheromonal group recognition mechanism leads to the maintenance of a stable social relationship among otherwise potential adversaries (264, 265, 398).

The pheromones that signify group membership may be released directly from the bodies of the animals or they may be deposited in the environment as scent marks. In most cases, it is difficult to differentiate whether the recognition of a social group is accomplished through the possession of a distinctive common odor—or the deposition of a common odor at communal scent marks—by all group members, or whether the recognition is accomplished on the basis of each animal having learned the separate individual odors of its group mates. Mechanisms do exist in some species for distributing a common odor among all members of the group. Rabbits (448) and tree shrews (403) deposit scent marks not only on various objects in their environment, but also upon each other, thus spreading their pheromones onto all group members. The scent-marking pheromone of the gliding phalanger is produced mainly by the dominant males of each group (577). The males use a specialized marking behavior that results in the pheromone being distributed all over the body surfaces of all the individuals of the group. In fact, young phalangers learn to recognize others as members of the common group before they learn to recognize them as individuals.

Social insects also use pheromones to differentiate whether others of the species belong to the same colony or to a foreign colony. The method whereby a common odor is obtained by all members of a colony is not completely understood and may vary among different species. Some of the odors are probably obtained from the various volatile chemicals produced within the colony such as from food, nest materials, and metabolic wastes. These odors may become adsorbed on the body surfaces of the colony members (76, 256). Also, distinctive colony odors may be the result of small colony-to-colony dietary differences (706) or, in some cases, they may be genetically determined and result from the common inheritance of all members of a colony from a single queen mother (695). The continuous acquisition of the odor from other colony members or from the nest environment is often of great importance. For instance, a foreign ant or bee might initially be attacked by members of a given colony, but if the foreigner is protected from attack for some time, it acquires the common odor and becomes adopted into the colony (306, 307, 711). Thus, beekeepers, when introducing a new queen into a honeybee colony, frequently keep her enclosed in a small cage in the hive for a few days, thereby protecting her from attack, until she obtains the appropriate odor.

4.4 Recognition of Home

A continuum exists, from the recognition by an individual of its own odor, to the recognition by an individual of the odor of its specific home or nest, to the recognition by an individual of the odor of the home range within which it conducts its daily activities. In some cases, all three phenomena might be identical. They may consist of the recognition by an animal of a scent that it had deposited previously in a certain place. In other cases, the odorous recognition by an animal of its nest or home or of its home range may be accomplished by the learning of scents that are not produced by itself but are produced by other living or dead organisms associated with that area and that differentiate that area from unfamiliar areas. The distinction between these two cases has received little study. Therefore, the reader should be aware that in some cases the odorous recognition of home, nest, or home range may not be accomplished by a true pheromone.

The use of odors by animals in their recognition of the particular site to which they return each day probably occurs throughout much of the animal kingdom. Many mammals that occupy specific dens or resting sites recognize those sites by odor, although there is little direct experimental evidence in this area. Most evidence is observational and relates to the especially intense scent-marking that is performed by the animals when they are near their resting sites. The use of the scent by the owner in its future recognition of its resting site is only one possible explanation for the intensified marking behavior. The scent deposited there could also have a role in deterring invasion of that site by other individuals of the same species or by individuals of other species.

An odor characteristic of the home or nest may not only allow recognition by the occupant, but may also be used to guide the occupant to the home from a distance. The odor may evaporate from the home site and form an aerial trail in a downwind or downstream direction, or the occupant itself may have deposited a terrestrial pheromone trail on the substratum during its previous excursions from the site. In either case, the animal may recognize the trail from its own home as being distinct from that of other members of the species and may follow the trail to its home (Chapter 5, Section 5.5).

Birds are generally not considered to use olfactory guidance mechanisms. However, Leach's petrel recognizes and steers toward the source of the odor coming from its own nest (244). These birds, like most shearwaters and petrels, return to their individual nest burrows at night. A petrel arrives near the entrance to its nest by flying upwind, following the odor trail emanating from the burrow. It plummets to the forest floor when a few meters downwind from the entrance and then walks the remaining distance, apparently still following its specific odor trail.

Limpets are small univalve molluscs which often have individual home or resting sites on rocks. A site usually consists of a depression formed by the limpet's shell wearing away the rock over a long period of time. The limpets make periodic feeding excursions away from their sites. Considerable evidence indicates that as they travel they deposit a pheromone trail on the substratum. They recognize their own trails on the homeward trip and follow them back to the home site (67, 151).

In some cases, whole populations of animals within a species recognize and orient toward a communal home site by sensing odorous cues. Certain fish species, including salmon and char, hatch in the headwaters of streams. At a particular stage of maturity, they migrate to the ocean, where they may stay for many months or years. They then return to reproduce in the headwaters of the same stream in which they hatched. It has been known for some time that chemicals that are typical of the water of the home stream are used in their olfactory orientation as the fish swim from the ocean, bypassing the incorrect branches of the stream. In the case of the char, *Salmo alpinus,* a pheromone may be involved. The char are divided among a number of populations, each with a specific home stream. Their olfactory receptors are more responsive to the pheromone released from fish belonging to their own population than to that from foreign populations. As only some of the char from each population migrate to the sea each summer, it is possible that pheromone released by the "stationary" fish—those remaining in their specific freshwater stream—produces a chemical trail in the water that enables the migrants to orient to the correct stream and population (175, 467).

Honeybees, as well as many other social insect species, recognize the entrance to their specific colony by odor. The odor, called "hive atmosphere" by LeComte (368), is distinct from one colony to the next. The odor not only enables identification of the correct home colony but it also seems to act as a stimulant causing the returning insects to enter the nest (123, 256, 368, 508). Hive atmosphere may consist largely of the same chemicals that impart a characteristic colony odor to the individual insects themselves (Chapter 4, Section 4.3). It may also be similar or identical in chemical composition to the "footprint substance" which workers of the honeybee and the colonial wasp, *Vespa vulgaris,* deposit on the substratum from their feet when walking near or through the entrance to their colony (123). The evaporating molecules of the footprint substance form an odor trail that orients other bees or wasps toward the entrance.

4.5 Recognition of Home Range

The home range of an animal is that area within which it normally confines its day-to-day activities. The home range, then, includes the location of the home or nest, and some redundancy of the concepts presented here, relative to those presented in the preceding section, is unavoidable.

A large body of circumstantial evidence indicates that the frequent deposition of scent marks by many mammals as they move around within their home range serves to maintain their familiarity with the area and perhaps enables them to identify the area with themselves (309, 352, 450, 452). The situation might be similar to that experienced by humans who feel insecure when in an unknown environment, but secure in a familiar environment. If the analogy holds, then the "security" may be accomplished when an animal identifies its earlier presence in an area by perceiving its own pheromone evaporating from previously deposited scent marks. Perhaps the greatly increased intensity of scent-marking exhibited by animals of certain species when first placed in a strange environment partly serves the purpose of conditioning the area with the familiar odor (179, 403, 450). This simple explanation is not completely adequate for most present-day mammals, in which scent-marking often serves a number of different behavioral roles. However, Kleinman (352) has proposed that marking behavior, at least in the Canidae, originated as a means of familiarizing and "reassuring" the animal when it entered a strange area, and that other functions such as the bringing together of the sexes or the maintenance of territory developed later.

The scent marks deposited within an animal's home range may also serve to label the habitat for the animal's own use in traveling from one important place to another (309, 724). The use of habitual paths and game trails is widespread among mammals. Scent is frequently deposited along the paths, sometimes by highly specialized methods.

Males of some loris species connect frequently visited areas such as their feeding places and sleeping sites with trails of urine (295, 583). The animals "urine wash," that is, they urinate onto one hand and then rub the urine onto the other hand and the feet; the scent trails are then deposited on the branches that they handle as they move through the forest. The trails presumably serve as olfactory orientation guides for the animals, enabling them to move accurately from area to area, even at night when visual orientation is impaired.

The black rhinoceros establishes dung piles at random over its home range (225). Any one pile is used by a number of individuals of both sexes. When a rhino approaches a pile, it sniffs the dung extensively. It may then sweep the pile with its anterior horn and shuffle through it with its feet. After defecating on the pile, it kicks at and scatters the feces with its hind legs. The fecal material on the feet establishes a scent trail which is passively deposited by the animal as it moves through the area and which may be followed later by the same or another rhino.

5

Aggregation Pheromone Behavior

The term "aggregation" as used here refers to the localization of one or more individuals in the vicinity of a pheromone source. That an aggregation could consist of one animal may appear to be contradictory. However, the fact that one individual arrived may be only a coincidence, and the behavioral reactions that caused the animal to become localized near the source may be identical to the reactions that would have caused thousands to become so localized.

"Aggregation" is preferred to the more commonly used term "attraction" because, unlike the latter term, aggregation does not infer the reactions which caused the animals to aggregate. Attraction implies an oriented approach toward the source from some distance (165), while the phenomenon of localization may be the result of a variety of other behavioral reactions in addition to or instead of attraction. For example, an aggregation pheromone might cause only an arrestment of locomotion, preventing an animal from moving away from a pheromone source once it had arrived there by chance (592).

One of the most common behaviors seen among pheromone-stimulated animals is their approach toward the odor source. The resulting aggregations serve a great variety of functions. Pheromones causing aggregation may be used in bringing others of the species to a source of food, to a suitable habitat that is to be colonized, to a sexual partner, or to a site of alarm where aggressive behavior will be displayed. Such pheromones also play an important role in the regulation of the activities of social insects and in the parent–young relationships of vertebrates. These and other types of pheromone-induced aggregation behavior are discussed in the following sections.

5.1 Exploitation of Sources of Food

The use of terrestrial trail pheromones in the foraging behavior of ants has been studied extensively. Depending on the species, the pheromone may be produced in a variety of glandular structures located on

Table 2. GLANDULAR SOURCES OF ANT TRAIL PHEROMONES[a]

Subfamily	Glandular organ producing the pheromone	Pheromone emitted from
Dolichoderinae	Pavan's gland	Abdominal sternum
Dorylinae	Hindgut	Anus
Formicinae	Hindgut	Anus
Myrmicinae	Poison gland	Sting
	Dufour's gland	Sting
	Metathoracic tibial glands	Probably tarsi
Ponerinae	Hindgut	Anus
	Poison gland	Sting

[a] Summarized from Gabba and Pavan (209) and Blum (71).

the abdomen or legs, and the secretion is deposited on the ground as the ant moves along (Table 2). Terrestrial trails deposited by ants moving from a food source toward their nest were named "recruitment trails" by Wilson (707). When the recruiting ant arrives in the nest, it may perform specialized behaviors that stimulate others to leave the nest and encounter the trail. Also, the odor of the trail pheromone itself may in some cases cause ants to emerge from the nest. Recruited ants then follow the terrestrial trail to the food source, grasp some of the food, and carry it back to the nest, following the original trail and often depositing additional trail pheromone themselves. This results in a continuous reinforcement of the trail as long as food remains at its terminus. Once the food is exhausted, returning ants do not deposit pheromone and the trail dies out.

The terrestrial trails of most army ants are deposited during their outward excursions from the colony centers to conduct raids for food, or during emigrations of entire colonies (73, 573–575, 664). These "exploratory trails" (707) thus differ from the recruitment trails previously discussed. Army ants typically advance in massive columns, with the ants at the forward margin of a column moving a short distance into unmarked territory while depositing pheromone and then hesitating momentarily while other ants advance beyond them. This leapfrog advancing situation is not the rule in all army ant species, however. In the genus Leptogenys, a scout ant which has discovered amphipods that are suitable as food is followed single file by colony members along a trail that it lays to the location of the prey (201).

Some ants construct more than one type of terrestrial trail. The harvester ant, Pogonomyrmex badius, deposits a trail with pheromone from its Dufour's gland even in the absence of food (282, 284). Thus, an array of trails is formed, radiating like spokes out from the central nest. The trails apparently are not used in recruiting sister workers to a food source, but instead are used by the ants when finding their way home. If an ant locates food, it follows a Dufour's gland trail home and at the same time deposits a trail with pheromone from its poison gland. The latter pheromone causes trail-following behavior of nest mates, which are thus recruited to the food.

A major component of the terrestrial trail pheromone of two *Atta* species, the Texas leaf-cutting ant (671) and *Atta cephalotes* (514), has been identified as methyl-4-methylpyrrole-2-carboxylate (4). A component of the trail pheromone of the pharaoh's ant also has been identified as 5-methyl-3-butyl octahydroindolizine* (5) (515).

4   5

Another behavior used by some ant species to recruit nest mates to a new food find is tandem running (281, 424). The ant that has discovered food returns to the nest and then moves back toward the food followed closely by a single ant. The follower ant keeps contact with the leader by perceiving a pheromone on the leader's abdomen, and only one ant can be recruited at a time. Moglich *et al.* (424) suggest that tandem running may be the evolutionary precursor of terrestrial-trail communication behavior in certain ant species.

Stingless bees of a number of species recruit colony mates to food sources by releasing into the air large quantities of pheromone from their mandibular glands. The pheromone forms an aerial trail that is followed by other workers to the food (337, 389, 390). Some stingless bees obtain their food by plundering the nests of related species (68, 75) and utilize pheromones in the process. The mandibular glands of *Lestrimelitta limao* contain both stereoisomers of citral [geranial (6) and neral (7)], in a 2:1 ratio. The first *L. limao* plunderers of a foreign nest are invariably killed by the nest defenders, resulting in the liberation of these compounds. The airborne pheromone recruits large numbers of additional plunderers, which raid the nest. The high concentrations of geranial and neral that are released have the additional property of completely disrupting the social organization of the defenders, which often abandon their nest.

* The stereochemistry of this compound has not been determined.

47

Worker honeybees sometimes mark objects in the environment with a pheromone that is produced in their Nassanoff gland and contains both isomers of citral (**6, 7**) (588), as well as geraniol (**8**), nerolic acid (**9**), and geranic acid (**10**) (77). Investigators have proposed that the pheromone has a function of designating rich sources of food and recruiting other worker bees by means of aerial trails which emanate from the scent marks (79, 215, 588). However, there is some difference of opinion as to the importance of the marking behavior in normal foraging. Free and Williams (205) propose that honeybees seldom mark the flowers from which pollen or nectar are collected, although they deposit marks near sources of water whose locations may need to be precisely communicated to other water-collecting bees.

Some stingless bees form pheromone trails between food sources and their nest that are partly aerial and partly terrestrial. A scout bee of *Trigona postica*, having collected nectar or pollen, flies from the flowers toward its nest, stopping at frequent intervals at objects such as branches and leaves and depositing at each spot a scent mark from her mandibular glands (389). After arriving at the nest, she flies back toward the food following her own trail, and followed closely by a group of recruited bees.

Termites recruit colony mates to food sources by depositing terrestrial trail pheromones from the sternal glands located on their abdomens (Fig. 15). In many species, a trail laid by a foraging worker while traveling away from the nest is reinforced by the same individual when it returns to the nest after discovering food. The trail then serves to guide other workers to the food (611, 612, 633, 634, 637). Workers of some species of the genus *Nasutitermes* construct pheromone trails between their nests located on the branches of trees and food sources on the forest floor (637). The trails are used for long periods of time and are completely enclosed by the workers, forming tunnel-like runways.

Examples of chemicals that have been identified as components of termite trail pheromones are *n*-caproic acid (**11**) in *Zootermopsis ne-*

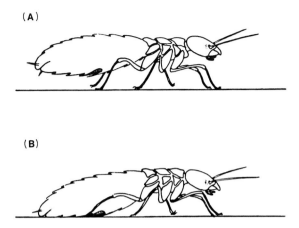

FIG. 15. Schematic representation of a nymph of the termite, *Zootermopsis nevadensis,* with the sternal gland (shaded area) elevated above the surface during normal activity (A) and contacting the surface during trail laying (B). [From Stuart (637).]

vadensis (291, 331) and 12-isopropyl-1,5,9-trimethylcyclotetradeca-1,5,9-triene* (**12**) in *Nasutitermes exitiosus* (59).

11

12

In marine snails, *Fasciolaria tulipa,* the aggregation pheromone is provided by the food source itself (615). The large snails are cannibalistic and are attracted to the odor of smaller snails, which they capture and eat. In turn, when the smaller snails detect the odor of nearby large individuals, they move away rapidly (Chapter 6, Section 6.1).

Pheromones are probably used extensively by many groups in addition to those dicussed above, to cause aggregations of species mates at sources of food. For instance, it is likely that the terrestrial pheromone trails produced by mammals are often followed to food sites (724); unfortunately, this use of pheromones by most animals is only surmised from observational evidence.

* The stereochemistry of this compound has not been determined.

5.2 Aggregation prior to Sexual Behavior

A remarkably diverse array of behaviors is stimulated by chemicals called sex pheromones. These pheromones are released into the environment by one sex and stimulate reactions in the opposite sex that either directly or indirectly enhance the likelihood that mating will occur. Indirect behavioral reactions are often involved with the approach of the responding animal toward the pheromone source and are considered here. Direct reactions, which entail courtship or copulatory responses by the pheromone-stimulated animal, are discussed in Chapter 8.2. In some species, a pheromone may cause only approach (and thus aggregation) of animals of the opposite sex; in other species, it may cause only courtship or copulation; and in still others, the pheromone may cause both types of reaction. In general, for both lower and higher animals, the sex that releases an aggregation-inducing sex pheromone is the female, and the sex that approaches the site of pheromone emission is the male.

When one considers single-celled animals, the distinction between "male" and "female" becomes difficult; however, considerable circumstantial evidence indicates that pheromones are often used to bring individual protozoans together for conjugation. The ciliate, *Rhabdostyla vernalis*, has two physiologically different forms (198). The smaller form, called a microconjugate, swims in a manner indicating that it orients to a pheromone gradient surrounding a macroconjugate, the larger form. Upon the arrival of the microconjugate at the macroconjugate, union occurs. Many similar cases are seen among the lower plants, in which pheromones secreted by female gametes attract motile male gametes. The first attracted male gamete of the marine brown alga, *Ectocarpus siliculosus*, becomes attached to the female gamete by the tip of its front flagellum. Once this fusion has occurred, the zygote loses its attraction for other aggregating male gametes, and they disperse (439). Other animal groups in which females are known to release sex pheromones that cause the aggregation of males are rotifers (222), nematodes (5, 138, 139, 190, 235–237, 241, 242, 317), arthropods (for partial reference lists, see 301, 592), fish (651, 661), snakes (218, 464, 465), and mammals (14, 39, 40, 129, 226, 227, 392, 440, 442, 445, 446, 565, 583).

In most cases, a male approaches a female mating partner by following an aerial odor trail (or the aquatic equivalent) to her vicinity. The pheromone may be released directly from the body of the female, or it may be released from scent marks deposited by her onto a substratum (440, 583). The urine scent marks of estrus female dogs and cats are well known for their ability to cause aggregation of males (565). Females of the Syrian golden hamster deposit secretions on surfaces against which they drag their vaginas; the odor of the scent marks causes males to approach from some distance (445).

Some female animals deposit terrestrial odor trails. During the reproductive season, female snakes of a number of species deposit a pheromone on the substratum over which they crawl (218, 464, 465). The males can follow the trails and thus arrive at the vicinity of the females. A variation on this theme is seen in certain spiders. The females incorporate a pheromone into the thread that they spin out and leave behind them as they move, and the males follow the silken pheromone trails to the females (174).

Tandem pairing behavior occurs in many termite species (637). Adults of both sexes leave the nest in a dispersal flight. After settling on the ground, they break off their wings. Then tandem pairs of a male and female are formed, with the male keeping contact with the female by a presumed pheromone emanating from her abdomen. The pair, led by the female, runs off to a suitable area and excavates the cell which will serve as the beginning of a new nest.

A female sex pheromone may also function as an arrestant, keeping males from leaving her vicinity after they have arrived there either by following a pheromone trail or by orienting to other, nonpheromonal cues (493). Some female mites release sex pheromone while in the quiescent deutonymph stage, before they molt to the sexually mature adult form. The pheromone stimulates males to approach from a distance and then acts as an arrestant, causing them to remain "in attendance" near her until she molts, at which time mating occurs (41, 148–150, 477, 478). A similar situation is seen in the crab hole mosquito (176). While in the pupal stage, the female mosquito releases a sex pheromone into the air through her spiracles which protrude above the water surface. Males are stimulated to alight on the water surface and then remain in attendance until the adult female emerges.

The opposite situation occurs in snails of the genus *Crepidula*. The free swimming, immature snails, not yet sexually differentiated, are attracted by the pheromone released from mature females (141). An attracted young snail attaches to a female and is then induced by a primer pheromone secreted by her to develop into a male, which mates with her. The young snails that do not attach to females eventually settle to the bottom and develop into females themselves.

In some animal species, females approach pheromone-releasing males; this occurs in certain nematodes (85, 241, 559), arthropods (for partial reference lists, see 301, 592), and mammals (129, 131, 164). Also *both* males and females of some nematode, arthropod, and mammalian species release pheromones that cause aggregation of the opposite sex (8, 11, 85, 130, 131, 163, 241, 410, 559, 616, 632).

Male bumble bees secrete pheromones that cause female aggregation (53, 203, 248–250, 364, 391, 625, 626). The males lay down semiaerial, semiterrestrial trails during their mating flights. They fly around a closed circuit, stopping occasionally to deposit pheromone from their mandibular glands at fixed points such as on branches of trees or blades of grass. A number of males often fly around and mark the same circuit. The females are stimulated to follow the same odor trails, thereby coming into contact with males.

5.3 Aggregation prior to Aggressive Behavior

The alarm pheromones constitute a group of chemicals which elicit a great variety of specific responses in the receiving individuals. The responses vary according to the species under consideration, the environmental context within which the pheromone is perceived, and the concentration of the pheromone. All of these pheromones are grouped in an "alarm" category by some authors, because in each case the secretion is released in response to a threatening situation, such as attack by an enemy. Animals of most species disperse from the area in which they detect an alarm pheromone (Chapter 6, Section 6.3); however, the response of social insects in some situations is an approach to the source of the alarm pheromone, followed by aggressive behavior. The approach phase is considered briefly here, and the role of the pheromones in stimulating aggression is discussed in Chapter 7, Section 7.1.

When an alarm pheromone is released in or near a hymenopteran nest, the nest mates typically are attracted to the pheromone source. The aggregation response is presumably adaptive in that it leads to defense of the nest (with its queen, food stores, and young) from enemy attack. However, when the insects are foraging away from the nest vicinity and perceive the same alarm pheromone, they typically flee. Again, the behavior appears adaptive, maximizing the likelihood that most of the insects will survive to return to the nest.

Workers of some ant species, when attracted to the source of alarm pheromone, display digging behavior (74, 155). The function of this behavior is not clear, but perhaps it serves to bury the source of the volatilizing pheromone, such as the crushed head of another worker, after the threatening situation has passed.

The alarm pheromone of many termite species is deposited as a terrestrial trail. The trail leads from the area that has received an enemy attack to other areas in the nest (635–637). Termite workers and soldiers follow the trail to the source of alarm where they perform behaviors such as attack upon an enemy or rebuilding of the walls, depending upon the type of stimuli they receive from the alarm source itself.

5.4 Maintenance of Societal and Family Groups

Complex insect societies may be composed of many thousands of individual animals that belong to a number of different castes and perform a number of different, specific duties (711). Communication is essential for coordination of the necessary interactions among the individuals, enabling the colony to function as an efficient unit. Much of the communication undoubtedly involves pheromones, but aside from obvious functions such as those dealing with mating (Chapter 8), aggression (Chapter 7), caring of the queen, and swarming, little is known of the role of pheromones in the every day life of the colony. Gross aggregation tendencies have been noted within the confines of a colony. For instance, individuals of the different castes often tend to aggregate together, and, in the termite, *Kalotermes flavicollis,* the aggregation has been shown to be caused by pheromones (676).

Care and feeding of the immature stages is probably often stimulated by pheromones. The tendency of ant workers to approach their larvae and to lick and groom them suggests that this intimate relationship is controlled by chemical stimuli emitted by the larvae. Both volatile (698), and nonvolatile, contact (697) pheromones have been demonstrated to cause ant workers to maintain close contact with their young. Other social aggregative functions such as food exchange among the adults and mutual grooming (705) are also probably controlled by pheromones, although these aspects of behavior have received little study.

Much of the regulation of a social insect colony is probably due to the releaser and primer pheromones produced by the queen. The best known of the chemicals involved is the "queen substance," 9-oxodec-*trans*-2-enoic acid (13), produced by the mandibular glands of the queen honeybee (16, 125). This chemical has important primer effects, such as the inhibition of ovary development in the workers (432, 472), as well as a variety of releaser effects, such as the attraction and sexual stimulation of drones when a virgin queen makes her nuptial flight from the colony (214). Within the colony, queen substance stimulates "retinue" behavior of the workers, causing them to approach the queen, to remain in attendance around her, and to feed and groom her (215, 585, 641). Other chemicals produced by the queen are probably also important in stimulating retinue behavior, since removal of her mandibular glands, and the subsequent cessation of production of 9-oxodec-*trans*-2-enoic acid, does not result in cessation of the tendency of workers to cluster around and tend her. Retinue behavior may serve a very important function in addition to the care of the queen in that it probably causes worker bees to maintain such close contact with her that they obtain sufficient queen substance to inhibit the development of their ovaries (432, 472). Retinue behavior, stimulated by pheromones produced by queens, occurs not only in bees, but also in ants (638, 698, 707, 708) and wasps (294, 296).

13   14

When honeybee colonies move from their parent nests to new locations, they form swarms consisting of a queen plus a large number of workers (711). A swarm usually alights and forms a cluster on some object, such as the branch of a tree, at least once during its migration to the new nest site. The location at which the cluster will form and the location of the final nest site is determined by advance scout bees that release Nassanoff pheromone (6–10) when they find a suitable place. A pheromone released by the queen, 9-hydroxydec-*trans*-2-enoic acid (14), acts as a behavioral stabilizer, causing the swarming bees to remain aggregated in a cluster if the queen is with them (124). If the queen is lost from the swarm, the worker bees usually find her by sensing the 9-oxodec-*trans*-2-enoic acid (13) that she releases (608), and the swarm moves to her location. Bees of a dequeened swarm can recognize their own colony queen from a foreign queen, apparently by perceiving the small differences in colony odors adsorbed on the queen's body (76, 430). The workers release Nassanoff pheromone when they find their own queen, stimulating the remainder of the swarming bees to form a cluster in that location (76, 122, 215, 408, 431, 675). On the other hand, if they perceive a foreign queen, the workers release alarm pheromone and attack her.

Aggregation behavior is also a vitally important aspect of the social life of vertebrate species, and the roles of pheromones in bringing interacting animals together for such activities as mating and colonization of habitats are considered in other sections of this chapter. An aspect which is considered here involves the role of pheromones in stimulating vertebrate parents and their young to remain together or to come together after periods of separation. Rat pups approach a lactating female in preference to one that is not lactating (376–378, 425, 426). The discrimination is due to a maternal pheromone released by the lactating female. She starts to produce the pheromone 14 days after the young are born and continues to emit it for an additional 14 days. As the pups become responsive to the pheromone when they are 14 days old, the pheromone seems to synchronize mother–young relationships during the later stages of lactation when the young begin to wander from the nest but must still return for nursing for an additional 2 weeks. Production of the pheromone is dependent on the female's titer of prolactin, the hormone necessary for milk secretion, plus stimuli received from nursing young that are at least 2 weeks old (426).

A similar maternal pheromone is important in causing the young of mice and Syrian golden hamsters to remain near a lactating mother (154, 166). Devor and Schneider (166) hypothesize that the hamster pheromone operates as an "olfactory tether," preventing the pups from straying dangerously far from the nest area during their early exploratory phase.

Adults of a number of fish species utilize olfactory stimuli in "caring" for their offspring. They respond to a pheromone released by the young by remaining in the area containing the chemically conditioned water. The aggregation response can be elicited by pheromone-treated water alone, even if the young fish are removed (363, 454).

5.5 Colonization of Habitats

Many animals, once they have settled in certain habitats, are physically incapable of moving to new locations. For these forms, pheromone signals from older individuals that are occupiers of the habitat may indicate its suitability and may cause aggregation of the still mobile, young animals in that area.

Sessile marine animals, such as barnacles and oysters, often occur in dense aggregations. The barnacle aggregations are caused by a pheromone released by adults that are attached to surfaces which presumably represent suitable habitats. Nearby, free-swimming barnacle larvae perceive the pheromone and are stimulated to approach the surfaces. If they then detect the pheromone as being adsorbed on the surfaces, the larvae settle there and develop into attached adult barnacles (156, 157).

In other animal species, the adults are free to select the habitat in which their offspring will be restricted. The habitat must be suitable for growth and development of the young; thus, females of many insect species are highly selective in their choice of locations in which they lay their eggs. The females of some mosquito species oviposit preferentially in water that has recently contained eggs, larvae, or pupae of the same species (159, 287, 325, 471). The behavior seems to be advantageous in that the habitat selected is one that has been previously demonstrated as suitable for survival of the young. Two pheromones are involved in the selection of an oviposition site by the mosquito, *Culex tarsalis*: one, produced by larvae and pupae, is detected by the female only upon contact with the conditioned water and induces her to lay eggs (287); the other, liberated into the water by the adult females during their egg laying activities, volatilizes from the water surface and causes other females to approach from a distance.

Females of both the sheep blowfly (34) and the desert locust (468, 469) gather in tight clusters with others of their species when laying eggs. The pheromones produced by the egg-laying females have a weak attraction effect, but a strong contact effect, maintaining the females in dense aggregations once they have come together and stimulating them to oviposit en masse. These insects, like mosquitoes, are opportunists in that they must locate the often scarce habitats which are suitable for development of their young.

Bark beetles and ambrosia beetles, although not living in highly organized societies, utilize efficient pheromone communication systems that result in large populations concentrating in suitable host trees. Depending on the species of beetle, the tree that is colonized may be healthy, weakened, dying, or dead. The beetles chew through the bark and establish tunnels in which they mate and lay their eggs. When the larvae hatch, they tunnel further through the xylem, phloem, or bark. The colonization process is sufficiently efficient that almost all available space and food resources are utilized by the massive numbers of beetles that aggregate in a given tree.

Colonization begins with the alightment on a suitable host tree of an initially invading beetle, which may be a male or female (but not both), depending on the species. Those species in which the initial invaders are male are typically polygamous, and several females will eventually be found in each tunnel with a single male. Species in which the initial invaders are females are monogamous, with single pairs of males and females eventually becoming established in each tunnel (91, 133–137, 182, 275, 399, 462, 463, 475, 476, 544, 547, 548, 551, 576, 681, 688, 689, 715, 716).

The initial invaders may use a number of orientation cues in their approach to an appropriate host, including the odor and visual image of the tree. In species that attack weakened trees, an initial invader bores its entry tunnel and then releases an aggregation pheromone with its feces. If the tree is too healthy—and thus an inappropriate host—it resists attack by secreting a copious flow of resin, resulting in the death of the invader before it can release the pheromone (485, 488, 682, 684, 718, 719, 726).

In species that attack healthy trees, an initial invader typically releases its aggregation pheromone before tunneling into the bark. This behavior seems to have adaptive value, because the likelihood that the initial invader will be killed by the tree's resin-defense mechanism is high. Many other invaders must be recruited to initiate tunnels until the tree becomes sufficiently weakened as a result of the mass attack that resin secretion diminishes and successful tunneling and colonization by the beetles can occur (684, 686, 690).

The aggregation pheromone secreted by the initial invaders, often in conjunction with resin odors from the tree itself, causes the recruitment of beetles of both sexes to the host. The behavior of the aggregating beetles differs according to sex. Those beetles that are of the same sex as the initial invaders bore new entry tunnels and release pheromone, thereby intensifying the pheromone concentration in the air near the tree and leading to increased aggregation of beetles from the surrounding environment. Beetles of the opposite sex may also release a pheromone when they arrive at the tree. These beetles are stimulated to enter tunnels made by beetles of the initially invading sex, whereupon mating occurs and eggs are laid. A number of complex behavioral responses that are not well understood occurs among the beetles of both sexes to ensure that the host is optimally colonized. It appears that the changing concentrations of pheromones and host-tree volatiles, plus the differing blends of chemical constituents in the pheromones secreted by the male and female beetles, often determine the numbers and sex ratios of the additional beetles that are recruited (42, 88, 207, 213, 234, 308, 353, 475, 476, 484, 486, 546, 683, 689, 716, 717).

Interchanges of stimuli between males and females assure that the two sexes get together in the tunnels. Often the opposite sex from that which initially bored a tunnel is arrested in its locomotion when it perceives the high pheromone concentration near a tunnel entrance (47, 308, 718, 719). The arrested beetle may also be stimulated by the high pheromone concentration to emit a characteristic sound that causes the occupier of the tunnel to permit it to enter. In addition, when it hears the sound, the tunnel occupier may release another pheromone into the air which is "antiaggregative," causing other flying beetles not to approach or alight in the area (387, 546, 549, 550, 552, 554, 555).

Several factors may operate separately or together to stop aggregation of additional beetles after the tree has become optimally colonized. While these factors differ from species to species and, in fact, for most species have not been studied, they probably include the following: (1) The host tree ceases its resin exudation as it becomes weakened through beetle attack. If the resin odor must be present in addition to the pheromone to stimulate aggregation, then further colonization ceases (504). (2) The concentration of the aggregation pheromone builds up in the air near the tree during the time when increasing numbers of beetles are arriving and initiating their own release of the secretion. The concentration may become so high that it causes the beetles that are still approaching from downwind to orient visually to nearby vertical objects before they arrive at the colonized tree. This phenomenon often results in intensive beetle attacks on trees close to a tree that is in advanced stages of colonization (212, 213, 685). (3) The quantities of the "antiaggregative" pheromone secreted by the initial invaders after potential mates arrive at the mouths of their tunnels may reach such high levels that the approach of additional beetles to the entire tree becomes inhibited (545). (4) The potential mates of the initial invaders may themselves secrete a pheromone after arriving at the tree that results in the inhibition of further aggregation (462, 463, 546, 555).

The chemical constituents that have been identified from certain of the bark beetle aggregation pheromones are shown in Table 3 and Fig. 16. Also included in Fig. 16 are two host-tree resins, a-pinene (25) and

Table 3. REPRESENTATIVE CHEMICALS THAT HAVE BEEN IMPLICATED AS PHEROMONE COMPONENTS OF CERTAIN BARK BEETLE SPECIES [a,b]

Pheromone component	Bark beetle species				
	California 5-spined ips	*Ips confusus*	Western pine beetle	Southern pine beetle	Douglas fir beetle
Ipsenol (15)	(605)	(725)			
Ipsdienol (16)	(605)	(725)			
Frontalin (17)			(343)	(343)	(487)
exo-Brevicomin (18)			(607)		
endo-Brevicomin (19)			(607)	(687)	
Verbenone (20)		(725)	(502)	(502)	(556)
cis-Verbenol (21)	(605)	(725)			
trans-Verbenol (22)		(725)	(502)	(502)	(553)
3-Methyl-2-cyclohexenone (23)					(344)
3-Methyl-2-cyclohexenol (24)					(344)

[a] The chemicals are not differentiated here as to the sex which produces them, and in some cases they may be produced by both sexes. Lightface numbers in parentheses indicate the reference to the chemical identification. Chemical structures of these pheromone components, as well as some host terpenes that may act with some of the components in causing beetle aggregation, are shown in Fig. 16. Boldface numbers in parentheses following components are the structure numbers as they appear in Fig. 16.

[b] Biological functions of many of these individual chemicals are not known. In some cases they may function in combination with other pheromone components and with host volatiles to cause aggregation of responding beetles. In other cases they may function to inhibit aggregation, and in still other cases, they may have both behavioral roles, depending on the context in which they are received.

myrcene (26), which interact with certain of the pheromone components to cause beetles of some species to aggregate. Further references to the chemistry of these secretions are given by Borden (89). The pheromones are complex, often consisting of three, four, or more components manufactured by one sex. In most cases, the exact roles of the individual components and the biological effects resulting from shifting ratios of the components are unknown. The compounds are mainly terpenoids, and their similarity to many of the chemicals that occur as a natural part of the host-tree resin is striking (Fig. 16). This similarity in chemical structures is probably not the result of coincidence; rather, increasing evidence indicates that in some cases the host terpenes may be metabolized within

FIG. 16. Representative chemicals that have been implicated as pheromone components of certain bark beetle species, plus host terpenes that interact with certain of the pheromone components to cause beetles of some species to aggregate. Beetles producing the pheromone components are indicated in Table 3.

Pheromone components

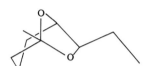

Ipsenol

15

Ipsdienol

16

Frontalin

17

exo-Brevicomin

18

endo-Brevicomin

19

Verbenone

20

cis-Verbenol

21

trans-Verbenol

22

3-Methyl-2-
cyclohexenone

23

3-Methyl-2-
cyclohexenol

24

Host tree volatiles

α-Pinene

25

Myrcene

26

the insects, resulting in the formation of certain of the pheromone components (Chapter 11) (288–290).

Animals that are freely mobile may still need to select suitable places as temporary habitats. Many animals which do not have permanent homes must rest in sheltered areas during their inactive times of day or during inactive seasons of the year. Often, the first individuals to settle in suitable resting or hibernating places communicate the location to others of the species by means of pheromones.

When ready to hibernate, animals of many species form large aggregations in protected areas. Certain coccinellid beetles and certain snakes locate their hibernation sites and are stimulated to remain at the sites after they have arrived there by sensing a pheromone emitted by other individuals (181, 218, 280, 465). The snakes, when proceeding toward a cavity to be used for hibernation, may follow terrestrial odor trails deposited by other snakes that preceded them.

Bedbugs and terrestrial isopods (wood lice) form resting aggregations during their inactive time of day. The first individuals to locate a suitable sheltered area release a pheromone that causes others of the species to aggregate there (206, 362, 384).

5.6   Other Aggregation Behavior

Other functions served by pheromone-induced aggregation behavior are so numerous that a complete listing is impractical. Therefore, a few selected examples are presented here to provide an understanding of the functional diversity involved.

During the feeding phase of their life cycle, cellular slime molds exist as independent single-celled amoebae, living in damp environments in the soil and feeding on bacteria. When the food supply is exhausted, some of the amoebae of *Dictyostelium discoideum* (the most studied species) emit pulses of a pheromone which has been identified as cyclic 3',5'-adenosine monophosphate (**27**) (361). Nearby amoebae respond by also emitting pulses of the pheromone and by moving toward the original signal source. In this way, each individual acts as a local signal source, causing other amoebae to move first toward their nearest neighbors. The aggregating individuals form streams which eventually coalesce into a "center" at the original signal source (Fig. 17) (82, 83, 516). The center then differentiates into a multicellular organism. Some cells form a stalk that rises into the air and others form a fruiting body at the apex of the stalk.

A pheromone released by males of the beetle, *Lycus loripes*, causes aggregation of beetles of both sexes. The beetles are distinctly colored and

27

release offensive chemicals when attacked by predators. The aggregations are presumably of survival benefit, because a predator, having had a distasteful experience with one beetle, will recognize the distinctive markings of the other, nearby individuals and tend to avoid them (185).

Recognition of the home or nest to which animals regularly return has been discussed in Chapter 4, Section 4.4. In most cases the animals not only use a characteristic odor to distinguish their own home from those of others of the species, but they also use the odor as a guide to enable orientation toward the home from a distance.

Fish of many species remain together in dense aggregations or schools. Although vision appears to be the primary sense involved in maintaining the individual fish in the correct spatial relationship with each other, some investigators have found that certain schooling fish also tend to approach water scented by others of the species (270, 333, 720). Possibly an aggregation pheromone operates in conjunction with vision in main-

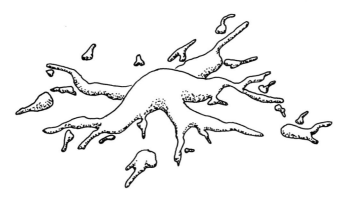

FIG. 17. Aggregation phase of the cellular slime mold, *Dictyostelium discoideum*, with streams of amoebae moving toward a central collection point. [Adapted from Bonner, J. T. (1963). How slime molds communicate. *Sci. Am.* **209**, 84–93. Copyright (1963) by Scientific American, Inc. All rights reserved.]

taining the schools during the day and then becomes the major schooling stimulus at night.

A pheromone, 2-methoxy-5-ethylphenol (28), is given off from the feces of the migratory locust, *Locusta migratoria migratorioides* (466). The chemical not only serves as a releaser pheromone, causing the young hoppers to aggregate, but also acts as a primer pheromone, inducing the physiological and morphological changes that result in the transformation of the hoppers to the migratory phase.

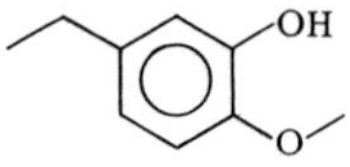

28

In some cases, pheromones are known to cause aggregations, but the biological functions of the aggregation behaviors are unclear. The aggregation pheromones of cockroaches are a case in point (44, 45, 108, 116, 297–300, 539). Depending on the species, the pheromone may be released from the feces or from various surfaces of the body, and it may cause aggregation of only nymphs or of all stages and sexes. The aggregation pheromone of the cockroach, *Blaberus craniifer*, has been identified as a mixture of chemicals, including undecane (29) and tetradecane (30).

29    30

A number of beetle species are characterized by a pheromone released by either males or females—depending on the species—that causes aggregation of both sexes. Some of these pheromones have distinct functions and were discussed earlier, in reference to the colonization of suitable hosts by bark beetles. In other species, however, including the boll weevil (96, 260, 423, 670), the Japanese beetle (229, 350, 351), *Pissodes* weevils (86), and certain anthicid and meloid beetles (1), the biological significance of the aggregation is not clear. The aggregation pheromone secreted by male boll weevils consists of a mixture of the following four chemicals (670):

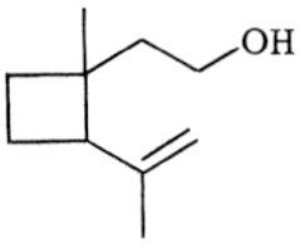

31   32

(+) – *cis*-2-isopropenyl-1-methylcyclobutaneethanol   *cis*-3,3-dimethyl-$\Delta^{1\beta}$-cyclohexaneethanol

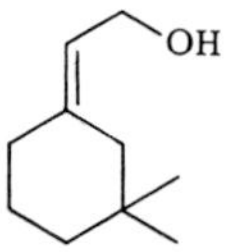

33

cis-3,3-dimethyl-$\Delta^{1\alpha}$-
cyclohexaneacetaldehyde

34

trans-3,3-dimethyl-$\Delta^{1\alpha}$-
cyclohexaneacetaldehyde

Many other animals, including planaria (409, 507); ticks (223, 224); "solitary" bees, *Augochiora pura* (18); and snails, *Littorina littorea* (169), produce pheromones that lead to aggregations of species mates. It is likely that these aggregations, as well as many of those discussed above, often serve a variety of functions, including the bringing of others to a suitable food source and the bringing together of the two sexes prior to mating.

6

Dispersion Pheromone Behavior

Although few detailed studies of mechanisms of dispersion have been conducted, the pheromones inducing this behavior must often stimulate locomotion and, simultaneously in some cases, cause taxis reactions that direct the animals away from the odor sources. In contrast, aggregation pheromones were generally found to result in taxis reactions directed toward the source and in arrestment of locomotion at high chemical concentrations.

Dispersion pheromones may serve a number of biological functions, such as maintaining an optimal separation between individual animals, maintaining separation between territorial social groups of animals, causing dispersion of conspecifics when a threatening situation is present, or countering the influence of aggregation stimuli when aggregation is inappropriate. These categories are somewhat artificial and tend to overlap.

6.1 Maintenance of Optimal Interindividual Spacing

As previously discussed (Chapter 5, Section 5.6), cellular slime molds are amoebae that aggregate at the end of their feeding period and form multicellular fruiting bodies. Both before and after aggregation, their behavior is characterized by mutual repellency. While grazing on bacteria, each individual amoeba releases a pheromone that causes others to keep their distance (82, 83), thus benefiting the species by maximizing the likelihood that the amoebae will locate and ingest all of the available food. Once the food is consumed and the amoebae have aggregated into the centers that will form fruiting bodies, a pheromone released from each center prevents other centers from forming nearby. This pheromone, which may be identical to that which earlier kept the individual amoebae dispersed, also acts to orient each rising fruiting body so that it is maximally displaced from adjacent bodies (84), by causing the stalk supporting the fruiting body to grow away from the area of highest pheromone concentration. Thus, if fruiting bodies are formed from

FIG. 18. Rising fruiting bodies of the cellular slime mold, *Dictyostelium discoideum*, release a pheromone that causes nearby bodies to lean away. [Adapted from Bonner, J. T. (1963). How slime molds communicate. *Sci. Am.* **209**, 84–93. Copyright (1963) by Scientific American, Inc. All rights reserved.]

nearby centers, the stalks lean away from one another (Fig. 18). Similarly, the stalk of a single fruiting body leans away from an adjacent vertical object by responding to its own pheromone concentration gradient, which is highest on the side toward that object (82). These spacing reactions of the centers and stalked fruiting bodies provide a maximum separation between individuals and allow the most uniform distribution of spores throughout the environment.

When populations of flour beetles (456, 458) and red flour beetles (457) are at high levels, the females distribute themselves uniformly throughout the medium in which they live, by dispersing in response to pheromones released by other females and by their larvae. This response, like that of slime mold amoebae, probably allows for the most efficient utilization of available food and permits the offspring to be deposited in less crowded areas.

When larvae of a number of lepidopterous species have completed their feeding and meet each other, they deposit a pheromone from their mandibular glands onto the substratum (152, 435), causing them to disperse. The effect is density dependent, in that the likelihood of the caterpillars meeting each other is dependent on the degree to which they are crowded. The dispersion enables the animals to enter areas of lower population density before they spin their cocoons, presumably increasing their chances of survival.

Large snails of *Fasciolaria tulipa* detect a pheromone released from smaller individuals of the same species and thereby are able to approach, capture, and devour them (Chapter 5, Section 5.1). In turn, the small snails can detect the odor of the larger ones and react by rapidly dispersing from the area (Fig. 19). Escape is made in a downstream direction or at right angles to the direction of water flow, with the escaping snails either rapidly gliding over the substratum or "leaping" by digging their

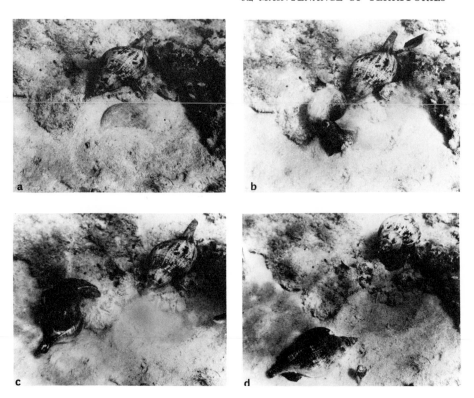

FIG. 19. Pheromone-mediated interactions among snails of *Fasciolaria tulipa.* (a) A snail (top) is placed directly upcurrent from another snail that is buried in the sand. (b, c) The buried snail responds by "leaping" out of the sand. (d) Regaining its footing, the snail glides rapidly downcurrent as the stimulus snail emerges from its shell and heads upcurrent. [From Snyder and Snyder (615).]

operculum into the sand and launching their shell through the water for a distance of a few centimeters (615).

6.2 Maintenance of Territories

A territory is an area that is occupied by one or more animals of a species, is recognized by them as their exclusive property, and is typically defended by them against invasion by other animals of the same species (402). The distinction is often vague between a fixed area constituting the territory and a certain volume of space surrounding individual animals as they move around and within which they will not tolerate others of the species. As pointed out by Jones and Nowell (312–316), mice form territories in defined parts of the habitat when their populalation density is low. At higher densities, mouse territoriality gives way to hierarchy formation, with the more dominant individuals having behavioral mechanisms that cause the withdrawal of subordinates from their immediate vicinity.

A variety of stimulus forms, including sounds, visual displays, and pheromones, may be used to advertise to others of the species that a territory is occupied. Pheromone scent-marking, commonly used by mammals, has the special advantage of advertising ownership even when the resident is absent (312). Mammals of many species scent-mark regularly as they move about within their territories, often intensifying the marking behavior when near the territory borders (173, 177, 184, 189, 233, 261, 263, 352, 360, 395, 403, 450, 490, 644, 650, 655, 656). However, as pointed out by Johnson (309), most evidence that the marks have a role in territorial defense is based on anecdotal reports, and a variety of other communicative functions might be served by the marking behavior.

The role of pheromone scent marks in territoriality has been most studied in mice and in the rabbit, *Oryctolagus cuniculus*. In both species, an individual that crosses the border between its own territory and a neighboring one has an immediate change in behavior, apparently induced by the territorial pheromone. The animal becomes alert and has an increased tendency to flee if it encounters an occupier of the territory (312–316, 397, 449, 450, 453). Some experimental evidence indicates that the pheromone odor alone may deter investigation of a marked foreign territory by male mice (316).

Social insects might also be considered as having territories, in that the nest and immediately surrounding area of one colony are defended against intruding members of other colonies. Intruders may recognize the territory and thereby alter their behavior when they detect a pheromone that is different from that of their own colony. Probably for this reason, ants frequently display escape behavior upon entering a foreign colony (696).

6.3 Dispersion in Response to Alarm Pheromones

Alarm pheromones released by a threatened or injured animal and causing the dispersion of nearby conspecifics are found in a wide variety of vertebrate and invertebrate species. Among the mammals, the alarm pheromone of mice has been the most intensively investigated (444, 543). A stressed mouse releases the pheromone with its urine, and the odor causes other to flee from the area.

Alarm pheromones of fish also have received considerable study (17, 481, 482, 578, 659, 692, 693). Von Frisch (692, 693) first observed that a school of minnows fled from its normal feeding place if the skin of one minnow was injured. Later workers found that the skin of most schooling cypriniform fish species contains alarm pheromones (481, 578). The olfactory sensitivity of the fish to the pheromone must be great; if a predator injures a minnow, the pheromone contained in only 0.01 mm² of the skin is sufficient to cause the entire school to flee. There is much variability in the responses of fish to alarm pheromones, but some generalizations can be made. Alarmed fish generally react as a school, which may become even more compact in response to the pheromone, and may drop to the bottom, approach the surface, or dash into cover, depending on the species and the environmental situation. At any rate, the school moves away from the location in which the pheromone was perceived (17, 659, 693).

Dispersion-inducing alarm pheromones have been demonstrated in a variety of invertebrates, including snails, sea urchins, earthworms, and insects. The pheromone of both snails and sea urchins, animals which often gather in dense groups, is released when one of the animals is injured by a predator. Most of the aquatic snail species studied by Snyder (613) respond to the pheromone released from a crushed conspecific by dropping to the bottom of the water and burrowing into the substratum. The main exception occurs in the air-breathing snail species, which react by crawling up, out of the water. A sea urchin of the species *Diadema antillarium*, when it detects the alarm pheromone from a crushed conspecific, moves rapidly away from the source, often mounting on its ventral spines and "racing" away for 1 or 2 m (Fig. 20) (614).

The earthworm, *Lumbricus terrestris*, when exposed to a predator, often releases a pheromone that leads to the rapid dispersion of other worms (497, 505). The pheromone is secreted in mucus, which also acts to repel the predator. This dual function of a chemical secretion, causing the simultaneous alarm of conspecifics and repulsion of the predator, is also seen in a variety of insect species, including aphids (455); the beetle, *Blaps sulcata* (332); bedbugs (385); the bug, *Dysdercus intermedius* (124); and ants (69, 404–406, 713; Chapter 5, Section 5.3).

When attacked by predators, aphids of many species produce droplets of fluid at the tips of the cornicles located on their abdomens. A pheromone, identified in several species as *trans-β*-farnesene (35) (94), volatil-

35

izes from the droplets, causing nearby aphids to move away or even to drop from the plant (Fig. 21) (346, 455).

Insect alarm pheromones often consist of mixtures of chemical components; two similar components have been identified in each of two

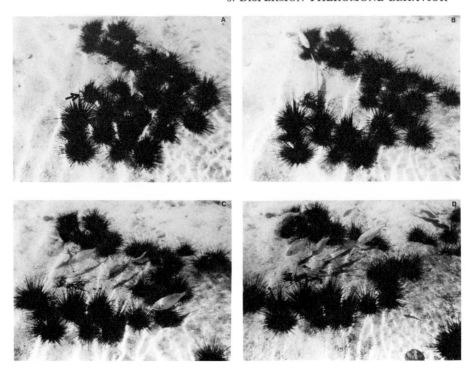

FIG. 20. Alarm pheromone responses by sea urchins, *Diadema antillarum*. (A) A clump of urchins immediately after one of them (arrow) is crushed. (B) Urchins downcurrent from the crushed urchin begin to move farther downcurrent. (C,D) Fish, mostly majarras and slippery dicks, rip apart the crushed urchin as downcurrent urchins continue to race downcurrent. Time span for A–D is approximately 2 minutes. [From Snyder, N., and Snyder, H. (1970). Alarm response of *Diadema antillarum*. *Science* **168,** 276–278. Copyright by American Association for the Advancement of Science, 1970.]

species of Hemiptera, the bedbug and *Dysdercus intermedius*. The chemical components produced by the bedbug are *trans*-2-octenal (**36**) and *trans*-2-hexenal (**37**) (385), while those produced by *Dysdercus intermedius* are hexanal (**38**) and *trans*-2-hexenal (**37**) (124). Similar compounds have been identified from the alarm pheromones of a number of ant species; however, since ant alarm has been studied most with respect to the aggregation and aggression behavior that is often induced in conspecifics near the nest, the chemistry of these pheromones is considered in Chapter 7, Section 7.1.

36   **37**   **38**

FIG. 21. A plant-feeding aphid, *Acyrthosiphon pisum,* under attack by a predaceous nabid bug. Droplets of secretion containing alarm pheromone have formed at the tips of the aphid's cornicles. (Courtesy of L. R. Nault.)

6.4 Antiaggregation

Females of many insect species aggregate near and lay their eggs in food material that will provide nourishment for the hatching larvae. When the density of females is high, this behavior may be disadvantageous in that so many eggs may be laid in each bit of food that the larvae will consume the material and die before they complete their development. Possibly resulting from the selective advantage of limiting the number of larvae to those which can be sustained by each bit of food, the females of certain species have developed a system for marking the food with a pheromone after laying their eggs. The pheromone prevents other females from laying their eggs in the same food and thus causes them to disperse. The end result is a more efficient utilization of all the available food and an increased chance for survival of the offspring.

Some hymenopterous parasites lay their eggs in other insects and have evolved pheromone-related behaviors to avoid multiple parasitization.

Females of the parasites, *Trichogramma evanescens* (560) and *Telenomus sphingis* (494), scent-mark the exterior of the insects in which they have oviposited. On the other hand, females of *Phaeogenes cynarae* inject the pheromone into the host insect when they deposit an egg therein (104). The pheromone can be detected by subsequently arriving females within 30 seconds after injection, causing them to disperse.

A pheromone-marking system is also used by the female apple maggot fly, which deposits her eggs in small fruits. After laying an egg, the female marks an area of the fruit surface that is proportional to the amount of food required for one larva to grow to maturity (Fig. 22) (492). The surface of a small fruit such as a cherry is fairly completely marked and thus receives only the original egg, whereas the surface of a larger fruit such as an apple is less completely marked by any one female and may receive proportionately more eggs.

A male of the yellow mealworm beetle, after approaching a pheromone-emitting female and mating with her, marks her with a pheromone, called an antiaphrodisiac by Happ (258), that prevents other males from being attracted to her. The females of this species are multiple maters, and the marking behavior is interpreted as increasing the likeli-

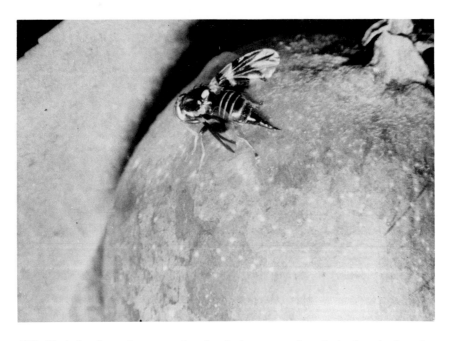

FIG. 22. A female apple maggot fly, after laying an egg in a fruit, deposits from her extended ovipositor a pheromone which deters other females from laying eggs in the marked area. (Courtesy of R. J. Prokopy.)

hood that a female will use sperm freshly transferred to her before another male mates with her.

Females of some animal species release dispersion-inducing pheromones when they are unresponsive for mating but are being pursued by one or more males. An unreceptive female of the ground beetle, *Pterostichus lucublandus,* when approached by a male usually runs from him. If pursued, she discharges a blast of liquid from the tip of her abdomen toward him, causing the male to stop running, to make cleaning movements with his legs on his face and antennae, and sometimes to become uncoordinated and roll over on his back. At the extreme, he may remain in a coma for several hours. Kirk and Dupraz (345) assume that this behavior is advantageous to the species when used by females during their egg-laying activities, because the male, which would eat the eggs, is immobilized while the female completes laying eggs, covers the egg chamber, and leaves the vicinity. A similar behavior is sometimes exhibited by a nonreceptive guinea pig female when she is pursued by a male. The female sprays urine, presumably containing an appropriate pheromone, onto the male, causing him to stop pursuit and to react by vigorous head shaking and sniffing and licking the urine (39).

Among the mammalian species (such as mice and rats) that establish dominance hierarchies, the pheromone odor of a dominant animal or of an area scent-marked by that animal is usually sufficient to keep subordinates at a distance. The pheromone of male mice is produced in the coagulating glands and is released with the urine (312, 313). The aversion of a subordinate to investigate a marked area is intensified if the subordinate has been recently defeated in an encounter with the dominant. A similar phenomenon occurs among yellow bullheads, fish which establish sophisticated social hierarchies (17). If one fish has established dominance, a subordinate flees when it perceives the dominant's pheromone.

Scent-marking by the black-tailed deer is part of an intricate communication system. Marking occurs particularly often near the sleeping site, which is the most important fixed point in the deer's home range. A strange deer that approaches the site often makes a sudden retreat when it perceives the scent (440, 441). In the same species, individuals approached by a strange deer frequently exhibit "rub-urinating," a procedure in which the hind legs are rubbed together when the animal urinates. Odorous chemicals in the urine presumably interact with chemicals produced by the tarsal glands and cause withdrawal of the strange deer (440).

Antiaggregation pheromones of bark beetles (see Chapter 5, Section 5.5) are sometimes released by individuals that are already located on suitable host trees. These pheromones inhibit other beetles from responding to the aggregation pheromones that also originate at those trees. The interactions of bark-beetle aggregation and antiaggregation pheromones are complex and vary from one species to another. When males of the Douglas fir beetle approach the mouth of the tunnel constructed by a female in response to her aggregation pheromone, they emit a characteristic sound. The sound causes the female to release the antiaggregative pheromone 3-methyl-2-cyclohexenone [Chapter 5, Section 5.5 (23)] which inhibits the in-flight approach response of additional beetles of both sexes (208, 344, 418, 545, 546, 554). In the related southern pine beetle, if more than one male arrives at the entrance to the female's tunnel, a fight ensues, with the males emitting rivalry sounds and releasing a blend of chemicals [including verbenone (20) and *endo*-brevicomin (19)] that has an antiaggregative function (503, 550). Apparently, some of the chemicals used in bark-beetle pheromones are multifunctional, stimulating aggregation when released at a low level by one sex and inhibiting aggregation when released at a higher level by the same or the opposite sex (550, 555, 556). Although the situation needs further clarification, the function of the interacting pheromones appears to be that of affording maximal colonization of a host tree, while at the same, preventing overcrowding.

7

Aggression Pheromone Behavior

Pheromones often play an important role in stimulating aggressive behavior among animals, with the aggression being directed either toward animals of other species or toward the pheromone emitter itself. On the other hand, certain pheromones may inhibit the aggressive tendencies of species mates. These aspects of pheromones as stimulators or inhibitors of aggression are considered in the following sections.

7.1 Stimulation of Aggression toward an Individual of Another Species

The alarm pheromones of social insects are used as signals to colony mates that a threatening situation is present. Near the nest, the individuals perceiving the pheromone usually approach the odor source (Chapter 5, Section 5.3) and display aggressive behavior. In most cases, the aggression seems to be triggered by an interaction of the pheromone stimulus with other appropriate stimuli coming from the intruding enemy. However, it has been demonstrated numerous times under laboratory conditions that the pheromone alone can sometimes cause aggressive behavior, without additional stimuli from the enemy. This explains why certain ants (191, 434, 501) and termites (428, 429) open their mandibles in an aggressive response when they perceive high concentrations of their species' pheromone.

The use of alarm pheromones as inciters of aggression against enemies has become highly specialized in the social Hymenoptera. The chemicals comprising the alarm pheromones are usually of low molecular weight and therefore rapidly volatilize, communicating the message to nearby individuals quickly. The chemicals frequently occur in mixtures and often originate in more than one glandular area on a single individual; however, the specific behavioral roles of each chemical in a mixture or of the pheromones from each glandular site are usually not well understood (12, 52, 78, 126, 501, 517, 587). Many aggression-promoting alarm-

Table 4. REPRESENTATIVE CHEMICALS THAT SERVE AS COMPONENTS OF THE ALARM PHEROMONES OF CERTAIN ANTS[a]

Ant subfamily	Genus	Chemical[b]
Dolichoderinae	*Iridomyrmex*	2-Heptanone (**41**)
	Dolichoderus	4-Methyl-2-hexanone (**39**)
	Tapinoma	2-Methyl-4-heptanone (**43**)
		6-Methyl-5-hepten-2-one (**44**)
Formicinae	*Acanthomyops*	Citronellal (**51**)
		Geranial (**52**)
		Neral (**53**)
		n-Undecane (**57**)
	Formica	Formic acid (**54**)
	Camponotus	*n*-Decane (**56**)
		n-Dodecane (**58**)
Myrmicinae	*Myrmica*	3-Octanone (**45**)
		3-Nonanone (**48**)
	Pogonomyrmex	4-Methyl-3-heptanone (**42**)
	Crematogaster	6-Methyl-3-octanone (**46**)
		2-Hexenal (**50**)
	Manica	4-Methyl-3-hexanone (**40**)
		4,6-Dimethyl-4-octen-3-one (**47**)
		3-Decanone (**49**)
	Messor	*p*-Xylene (**55**)
	Mycocepurus	*o*-Aminoacetophenone (**61**)
Ponerinae	*Paltothyreus*	Dimethyl disulfide (**59**)
		Dimethyl trisulfide (**60**)
	Odontomachus	2,5-Dimethyl-3-isopentylpyrazine (**62**)
		2,6-Dimethyl-3-*n*-pentylpyrazine (**63**)
		2,6-Dimethyl-3-*n*-butylpyrazine (**64**)

[a] See Blum (71) for references.
[b] Boldface numbers in parentheses following chemical identification are the structure numbers as they appear in Fig. 23.

pheromone chemicals have been identified from ants, and representative structures are shown in Table 4 and Fig. 23.

The same chemical may be used as part of the alarm-pheromone blend of a number of species, resulting in a considerable lack of species specificity; thus, the pheromone released by workers of one species may excite aggressive behavior in related species. However, because the approach and aggressive behavior are only stimulated in other ants that are within a few centimeters of the pheromone source and because this behavior is usually stimulated in or near the nest, only the insects belonging to the appropriate colony are available to respond, making the lack of specificity of little importance in nature (52, 70, 71).

The alarm pheromones of social Hymenoptera are produced in glands

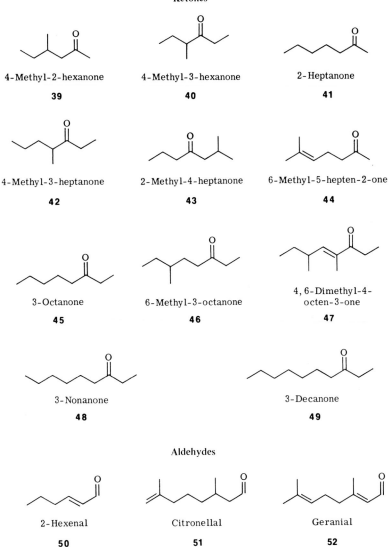

Ketones

4-Methyl-2-hexanone
39

4-Methyl-3-hexanone
40

2-Heptanone
41

4-Methyl-3-heptanone
42

2-Methyl-4-heptanone
43

6-Methyl-5-hepten-2-one
44

3-Octanone
45

6-Methyl-3-octanone
46

4,6-Dimethyl-4-
octen-3-one
47

3-Nonanone
48

3-Decanone
49

Aldehydes

2-Hexenal
50

Citronellal
51

Geranial
52

Neral
53

FIG. 23. Representative chemicals that serve as components of the alarm pheromones of certain ants (see Table 4). After Blum (71).

Acid

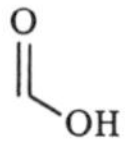

Formic acid

54

Hydrocarbons

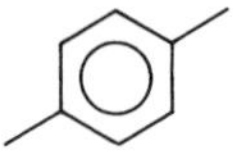

p-Xylene

55

n-Decane

56

n-Undecane

57

n-Dodecane

58

Miscellaneous

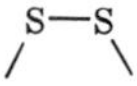

Dimethyl disulfide

59

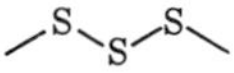

Dimethyl trisulfide

60

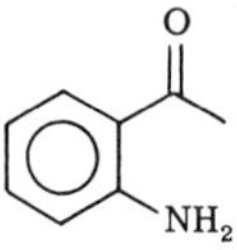

o-Aminoacetophenone

61

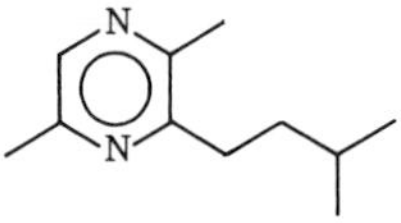

2, 5-Dimethyl-3-
isopentylpyrazine

62

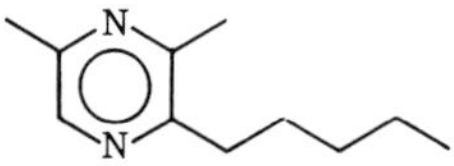

2, 6-Dimethyl-3-
n-pentylpyrazine

63

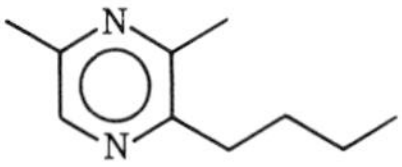

2, 6-Dimethyl-3-
n-butylpyrazine

64

FIG. 23. Continued.

that are closely associated with the organs of aggression: the mandibles or sting (126, 404–406, 517, 713). The pheromone may be released into the air when the insect is crushed by a predator, or it may be released through direct motor control by the threatened insect. Some species can

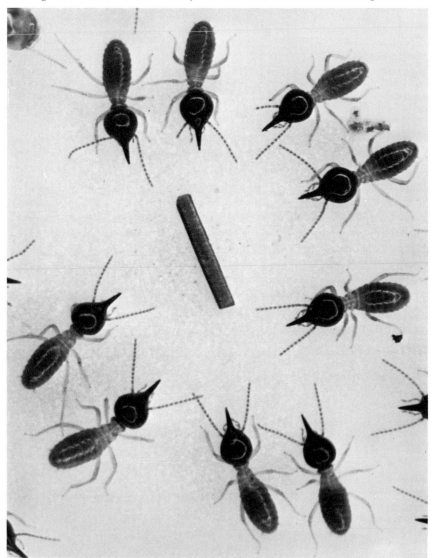

FIG. 24. Response of termite soldiers, *Nasutitermes exitiosus*, to the secretion sprayed by one of them from its cephalic nozzle onto a small metal bar. The secretion is a chemical irritant and entangling agent that effectively incapacitates the "enemy," and it also serves as an alarm pheromone that attracts other soldiers and causes them to be ready to spray their own secretion onto the target. (Courtesy of Thomas Eisner.)

control the direction of pheromone emission, as is the case with worker ants of the genus *Formica* which can spray a mixture of the contents of their Dufour's gland and poison gland directly onto an enemy (52).

A honeybee, when disturbed at the hive, opens its sting chamber and releases a pheromone which attracts other bees and stimulates them to attack dark, moving objects in the vicinity (78, 404–406). If the bee stings an enemy, it frequently grasps the skin or hair with its mandibles and releases from its mandibular glands an additional pheromone which also stimulates aggregation and aggression by other bees. These phero-mones may be responsible for the phenomenon familiar to bee keepers that more than one bee often is attracted to and stings an enemy in the same spot on its body (78, 204, 587).

Most termite alarm pheromones are emitted in the form of terrestrial trails that guide colony mates to a threatened area in the nest (Chapter 5, Section 5.3). However, in some termite species the pheromones are released into the air at the site of danger and directly stimulate aggres-sive behavior by nearby termites. Soldiers of *Nasutitermes* spp. have the front portion of their heads drawn out anteriorly with a hole at the tip, forming a nozzle through which they can expel a secretion containing a variety of compounds including α-pinene (**65**) at intruders (428, 637). The secretion not only immobilizes the intruders but also excites aggres-sive behavior among other soldiers (Fig. 24). A similar chemical, limonene (**66**), which has been identified from the alarm pheromone of *Drepanotermes rubriceps*, also is released by an individual at the site of danger and causes attack behavior by conspecifics (428, 429).

65   **66**

7.2 Stimulation of Aggression toward a Conspecific

Sometimes the odor released from one individual causes other mem-bers of the same species to approach and attack that individual. This is the case especially in dominance hierarchies, in which the more domi-nant animals may be stimulated to aggression when they perceive the odor of subordinates. It also occurs in relation to territoriality, with animals that coexist as a social group being stimulated to attack or threaten an animal having an odor foreign to the group. Such attack or threat behavior directed toward another animal, or sometimes toward an object scent-marked by another animal, occurs in a wide variety of mammalian species (39, 309, 440, 443, 449, 577, 660). The phenomenon has been especially studied in mice. Male mice are stimulated to attack strange males, and females are stimulated to attack strange females when they perceive the odor of the strangers (264, 266, 398, 535–538). The production of the aggression-promoting pheromones by male and female mice is related to, and possibly controlled by, the titers of male and female sex hormones, respectively (266, 370–373, 398, 436, 438). Most studies of mouse aggression-promoting pheromones have been concentrated on the chemicals that the animals secrete with their urine; however, male mice have an additional aggression-promoting pheromone that is produced in their preputial glands and is apparently liberated to the exterior directly, without being mixed with the urine (437).

Some fish attack others of the same species when they perceive appropriate pheromones. Yellow bullheads form sophisticated social systems within established territories, and a group of fish existing together within a territory have a strict hierarchical relationship to one another. A dominant fish is stimulated to attack when it perceives the odor of familiar subordinates in or near its hiding spot. However, if a strange fish enters the territory, the dominant fish, which can differentiate between the odors of the stranger and the familiar subordinates, may tolerate the subordinates in its hiding spot and attack the intruder (Fig. 25). The dominant fish then returns to its spot and evicts the subordinates. This pheromone communication among members of the hierarchy is very sensitive to social changes. For example, if a dominant fish is taken from its territory and stressed, such as by losing a fight with another fish, the change in status is chemically communicated to the other bullheads which will now be stimulated to attack the previously dominant individual (17, 663).

Social insects recognize the odor of foreigners entering their colonies

FIG. 25. A dominant yellow bullhead tolerating a subordinate fish in its hiding spot when it detects the presence of another, strange fish. (Courtesy of J. H. Todd and J. E. Bardach.)

and are thereby stimulated to attack them. The basis for this differentiation between nest mates and intruders appears to be odors that are characteristic for each colony (Chapter 4, Section 4.3). Similarly, when swarming honeybees have lost their queen and subsequently find a queen, they can differentiate by odor whether she belongs to their colony or to a foreign colony and accept or attack her accordingly (Chapter 5, Section 5.4). More than one queen cannot coexist within a single, established honeybee colony. If two are present, they recognize each other by odor and are caused to exhibit fighting behavior, resulting in the death of one of them (513).

Aggression-inducing pheromones are released by some male bark beetles during their colonization of suitable host trees (Chapter 5, Section 5.5). In certain species, if two males arrive simultaneously at the entrance to a tunnel in response to an aggregation pheromone given off from a female therein, they liberate a pheromone which causes typical male "rivalry" behavior, including the production of characteristic sounds and direct fighting (546, 554).

7.3 Inhibition of Aggression toward a Conspecific

Animals could not live together in organized societies if their aggressive instincts were not sometimes suppressed. For example, although males often act aggressively toward other males, it would be maladaptive for them to attack females. Male-to-female aggression in some mammalian species is inhibited by a pheromone released by the females with their urine (170, 437, 445). The pheromone is so effective that male mice occupying a territory will not attack a strange male if female urine is first rubbed into the stranger's fur. Instead, the resident males respond with "social investigation" as well as sexual behavior directed toward the stranger (170). Males of the fish species, *Bathygobius soporator*, also normally attack other males, but they are inhibited from displaying aggression if they smell the odor of a female (651). In some animal species, the secretion that acts to inhibit male aggressiveness might be that which also functions as the female sex pheromone.

When a number of adult male animals must coexist in a social group, the aggressiveness that they would normally display toward each other must be inhibited. A pheromone produced in glands associated with the genital tract of male mice produces a pheromone that inhibits aggression and leads to maintenance of a social relationship among males that have cohabited with each other for some time (264, 265). Also, the aggressive behavior between two strange male mice is reduced if one of them is rubbed with the urine of a male already known to the other (398).

Aggressiveness of adult mammals toward their young must also be inhibited. The role of pheromones in this important aspect of maintaining social order has received little attention; however, pheromones, or the lack of them, are probably often involved. It is possible that the young are protected from aggression because they lack the high titer of sex hormones that are apparently needed for synthesis of aggression-promoting pheromones.

Young worker ants may also receive protection against attack. When young *Formica* workers enter other colonies of the same species, they are typically adopted, whereas older workers are attacked and killed. The different response is caused by a pheromone found only in young workers that suppresses the normal aggressiveness of old workers toward individuals from foreign colonies (306, 307).

Aggression-inhibiting pheromones are used by some social insects to neutralize the effects of alarm pheromones. An ant that releases a pheromone in the presence of an enemy might itself be subject to attack by its sister workers. However, a chemical, 3-octanone (45, Fig. 23), which is released with alarm pheromone by workers of the ant, *Myrmica rubra*, inhibits the aggressive tendencies of other workers (668). The chemical might be deployed in such a way that the aggressiveness of the other workers is directed toward the enemy but not toward the pheromone-releasing ant.

7.4 Release of Pheromone as an Aggressive Act

The release of a pheromone to inhibit aggression in another animal may in itself constitute an aggressive act. Thus, in territorial animals, the marking of the territory with scent may operate to establish a threat to potential invaders, which are not only discouraged from investigating the marked area, but which are also intimidated so that if a fight were to take place, the territory occupier would most likely win (Fig. 26) (309, 312–316, 449).

Even in nonterritorial species, when two strange mammals come together, they may release pheromone directly from glands or scent-mark nearby objects, often as a prelude to or an actual part of a conflict between the animals (113, 187, 219, 440, 443). Scent-marking by male mammals, in particular, may become greatly intensified during aggressive encounters. Fighting male guinea pigs often spray urine containing a supposed pheromone directly onto each other (39); however, the behavioral effects of the pheromones employed during the conflicts among guinea pigs, as well as among other mammals, are poorly understood (496).

The "stink fights" between confronting males of the lemur (*Lemur*

83

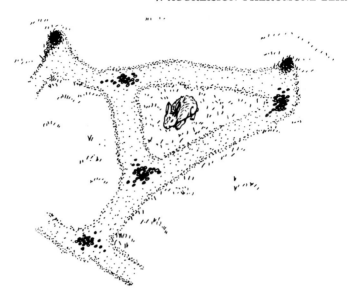

FIG. 26. Portion of a rabbit territory with intersecting runways leading from the entrances to the warren. Small mounds of fecal pellets are deposited by the rabbits at intervals along the trails, and pheromones that mark the pellets may inhibit the aggressive tendencies of rabbits that are foreign to the territory. [Adapted from Mykytowycz, R. (1968). Territorial marking by rabbits. *Sci. Am.* **218,** 116–126. Copyright (1968) by Scientific American, Inc. All right reserved.]

catta), which involve elaborate behaviors for pheromone-marking of the animal's own body as well as nearby objects, have received particular attention (189, 310, 311). The tail is marked by being pulled across specialized wrist glands and is then waved at the enemy. Also, branches are marked in sequence by a pheromone on the lemurs' hands as one of the males advances. First one male marks a branch and then the other, with pauses between. When the more aggressive male moves forward, it marks the same branches that its adversary had marked.

8

Sex Pheromone Behavior

Sex pheromones are chemicals that are secreted by animals of one sex and cause behavioral reactions in the opposite sex that facilitate mating. Because of their obvious role in attracting potential mates from a distance and their possible utility for manipulating the behavior of pest animals, sex pheromones have been studied extensively. In fact, probably half of the world's pheromone literature relates to the chemicals that are released by female moths and that stimulate the approach of males prior to mating.

One or more of a variety of reactions may be induced in the animal perceiving the pheromone, depending on the species. Most of the reactions fall into two categories: those that lead to aggregation near the pheromone source and those that are involved in close-range courtship or copulatory behavior. Another type of reaction that has received little attention is the synchronization of the time of sexual activity of the two sexes. Males of the ant, *Camponotus herculeanus*, when emerging from the nest and embarking on their nuptial flight, release a pheromone from their mandibular glands. The pheromone causes the females to take flight also, ensuring that the two sexes will be in the same environment at the same time (283).

8.1 Stimulation of Aggregation

The widespread use of pheromones to stimulate aggregation of potential mating partners was discussed in Chapter 5, Section 5.2. A number of the sex pheromones that are released by females and that cause the approach of males have been characterized chemically. Essentially all of these chemicals have been identified from female insects, and representative chemical structures are given in Tables 5 and 6 and Figs. 27 and 28. Many of the same chemicals also stimulate males to perform close-range courtship or copulatory behaviors once they arrive near the female.

Table 5. SEX PHEROMONE CHEMICALS IDENTIFIED FROM FEMALE IN-
SECTS OF THE ORDERS COLEOPTERA, DIPTERA, AND HYMENOPTERA,
CAUSING AGGREGATION OF MALES PRIOR TO MATING

Species producing the pheromone	Chemical[a]	Reference
Coleoptera		
Sugar beet wireworm	Valeric acid (**67**)	302
Black carpet beetle	*trans*-3,*cis*-5-Tetradecadienoic acid (**69**)	606
Trogoderma inclusum	(−)-14-Methyl-*cis*-8-hexadecenol (**72**), (−)-methyl-14-methyl-*cis*-8-hexadecenoate (**73**)	518
Grass grub beetle	Phenol (**71**)	273
Diptera		
Housefly	*cis*-9-Tricosene (**70**)	128
Hymenoptera		
Honeybee	9-Oxodec-*trans*-2-enoic acid (**68**)	214

[a] Boldface numbers in parentheses following the chemical name are the structure numbers in Fig. 27.

Table 6. SEX PHEROMONE CHEMICALS IDENTIFIED FROM FEMALE LEPI-
DOPTERA, CAUSING AGGREGATION OF MALES PRIOR TO MATING

Species producing the pheromone	Chemical[a]	Reference
Aegeriidae		
Lesser peach tree borer	*trans*-3,*cis*-13-Octadecadienyl acetate (**93**)	674
Peach tree borer	*cis*-3,*cis*-13-Octadecadienyl acetate (**92**)	674
Arctiidae		
Tiger moths (8 species)	2-Methyl heptadecane (**97**)	520
Bombycidae		
Silkmoth	*trans*-10,*cis*-12-Hexadecadienol (**76**)	117
Eucosmidae		
False codling moth	*trans*-7-Dodecenyl acetate (**79**)	499
Gelechiidae		
Angoumois grain moth	*cis*-7-*trans*-11-Hexadecadienyl acetate (**91**)	679
Pink bollworm	*cis*-7,*cis*-11-Hexadecadienyl acetate (**90**), *cis*-7,*trans*-11-Hexadecadienyl acetate (**91**)	292
Lymantriidae		
Gypsy moth	*cis*-7,8-epoxy-2-Methyl octadecane (**98**)	58
Noctuidae		
Cotton bollworm	*cis*-11-Hexadecenal (**96**)	531
Tobacco budworm	*cis*-9-Tetradecenal (**94**), *cis*-11-Hexadecenal (**96**)	531
Cabbage looper	} *cis*-7-Dodecenyl acetate (**78**)	48
Soybean looper		672
Fall armyworm	*cis*-9-Tetradecenyl acetate (**85**)	584
Beet armyworm	*cis*-9-*trans*-12-Tetradecadienyl acetate (**89**)	99

86

Table 6.   *Continued*

Species producing the pheromone	Chemical[a]	Reference
Southern armyworm	*cis*-9-Tetradecenyl acetate (**85**), *cis*-9,*trans*-12-tetradecadienyl acetate (**89**)	304, 500
Red bollworm	11-Dodecenyl acetate (**83**), *trans*-9,11-dodecadienyl acetate (**84**)	459
Spodoptera litura	⎱ *cis*-9,*trans*-11-Tetradecadienyl acetate (**88**),	649
Spodoptera littoralis	⎰ *cis*-9,*trans*-12-tetradecadienyl acetate (**89**)	459, 460 646
Pyralidae		
Raisin moth		98
Indian-meal moth		103, 160, 366
Mediterranean flour moth	*cis*-9,*trans*-12-Tetradecadienyl acetate (**89**)	102, 160, 365
Tobacco moth		100
Almond moth	*cis*-9-Tetradecenyl acetate (**85**), *cis*-9,*trans*-12-tetradecadienyl acetate (**89**)	97, 103, 366
European corn borer	*cis*-11-Tetradecenyl acetate (**86**), *trans*-11-tetradecenyl acetate (**87**)	355, 357, 529
Saturniidae		
Pine emperor moth	*cis*-5-Decenyl isovalerate (**99**)	271
Tortricidae		
Oriental fruit moth	*cis*-8-Dodecenyl acetate (**80**)	526
Grape berry moth	*cis*-9-Dodecenyl acetate (**81**)	528
European pine shoot moth	*trans*-9-Dodecenyl acetate (**82**)	610
Oblique-banded leaf roller	*cis*-11-Tetradecenyl acetate (**86**)	525
Summer fruit tortrix	⎧ *cis*-9-Tetradecenyl acetate (**85**),	421, 648
Clepsis spectrana	⎨ *cis*-11-tetradecenyl acetate (**86**)	421
Smaller tea tortrix	⎩	647
Fruit tree tortrix	⎱ *cis*-11-Tetradecenyl acetate (**86**),	480
Red-banded leaf roller	⎰ *trans*-11-tetradecenyl acetate (**87**)	357, 519
Fruit tree leaf roller	Dodecyl acetate (**77**), *cis*-11-tetradecenyl acetate (**86**), *trans*-11-tetradecenyl acetate (**87**)	530
Tufted apple bud moth	*trans*-11-Tetradecenyl acetate (**87**), *trans*-11-tetradecenol (**75**)	279
Codling moth	*trans*-8,*trans*-10-Dodecadienol (**74**)	56, 527
Eastern spruce budworm	*trans*-11-Tetradecenal (**95**)	699

[a] Boldface numbers in parentheses following the chemical name are the structure numbers in Fig. 28.

Our knowledge concerning the total complex of chemicals that constitute each supposedly identified pheromone is fragmentary. In early studies, a single chemical that was considered to constitute the entire pheromone was isolated and identified. However, as more detailed in-

Acids

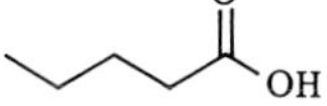

Valeric acid

67

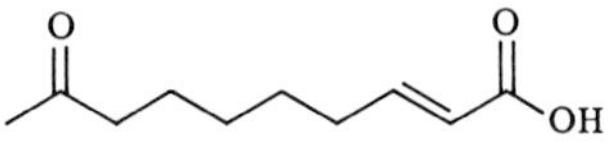

9-Oxodec-*trans*-
2-enoic acid

68

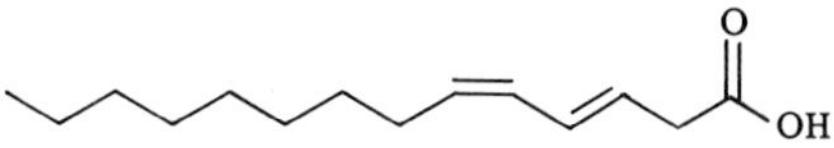

trans-3, *cis*-5-Tetra-
decadienoic acid

69

Hydrocarbon

cis-9-Tricosene

70

Alcohols

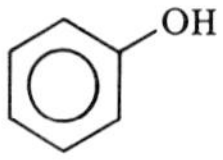

Phenol

71

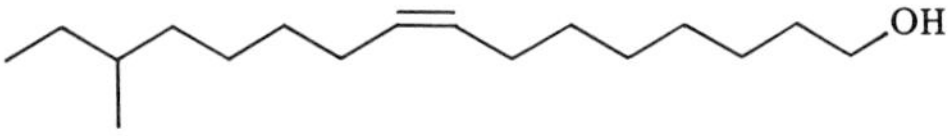

(−)-14-Methyl-*cis*-
8-hexadecenol

72

Ester

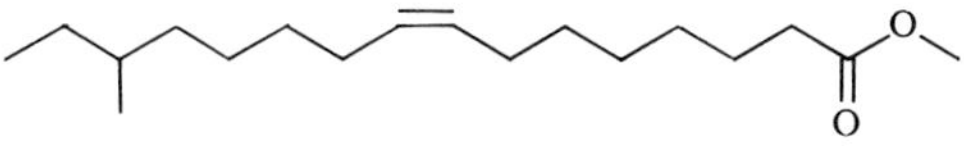

(−)-Methyl-14-methyl-
cis-8-hexadecenoate

73

FIG. 27. Sex pheromone chemicals identified from female insects of the orders Coleoptera and Hymenoptera (see Table 5).

Alcohols

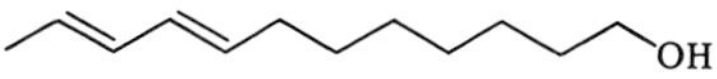

trans-8, *trans*-10-Dodecadienol

74

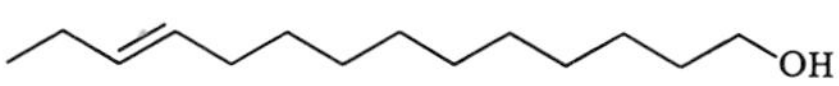

trans-11-Tetradecenol

75

trans-10, *cis*-12-Hexadecadienol

76

Acetate esters

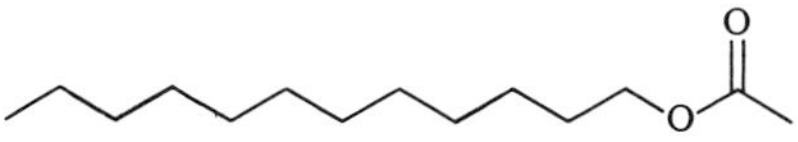

Dodecyl acetate

77

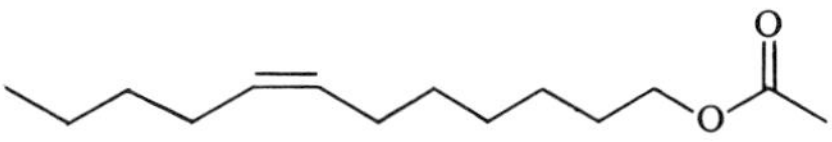

cis-7-Dodecenyl acetate

78

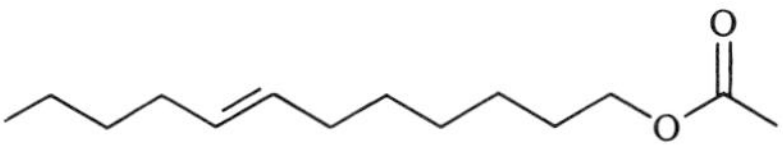

trans-7-Dodecenyl acetate

79

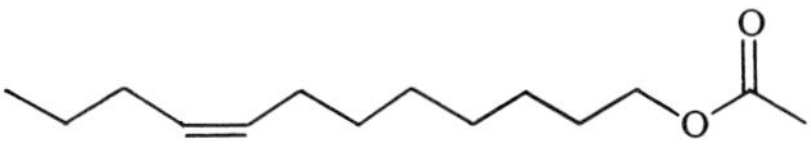

cis-8-Dodecenyl acetate

80

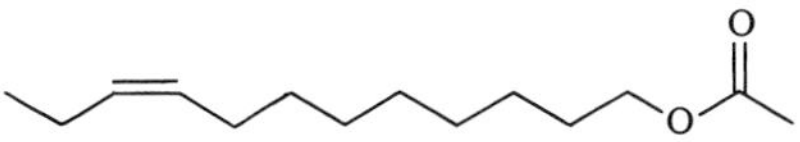

cis-9-Dodecenyl acetate

81

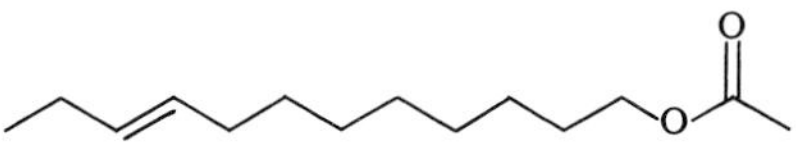

trans-9-Dodecenyl acetate

82

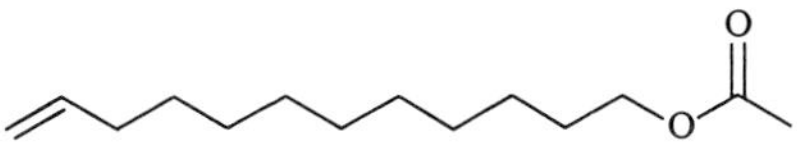

11-Dodecenyl acetate

83

FIG. 28. Sex pheromone chemicals identified from female Lepidoptera (see Table 6).

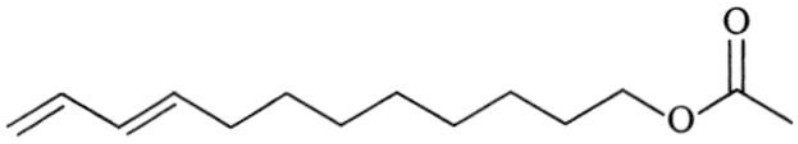

trans-9, 11-Dodecadienyl acetate

84

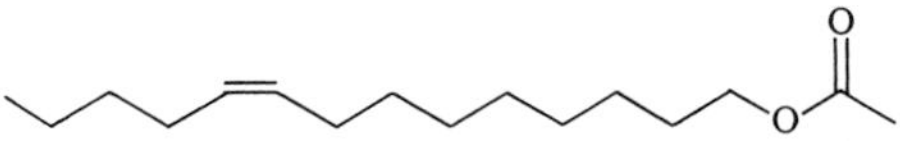

cis-9-Tetradecenyl acetate

85

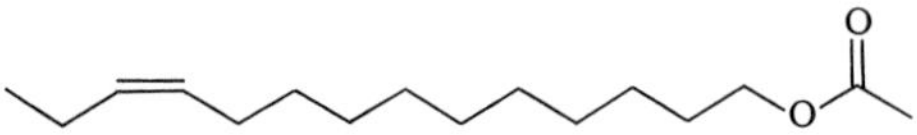

cis-11-Tetradecenyl acetate

86

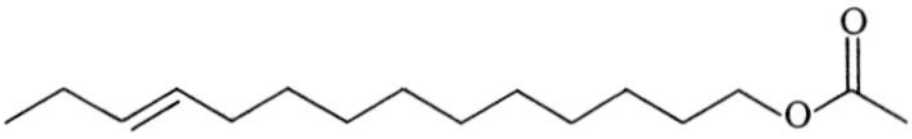

trans-11-Tetradecenyl acetate

87

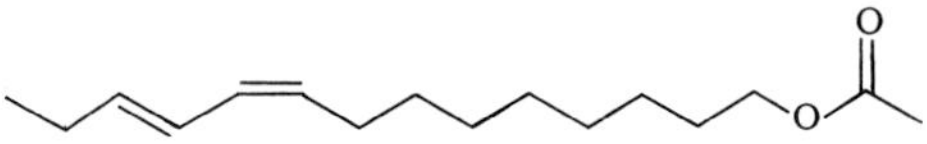

cis-9, *trans*-11-Tetradecadienyl acetate

88

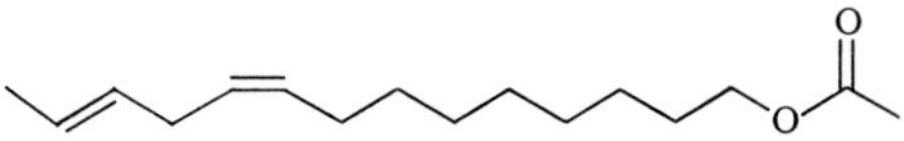

cis-9, *trans*-12-Tetradecadienyl acetate

89

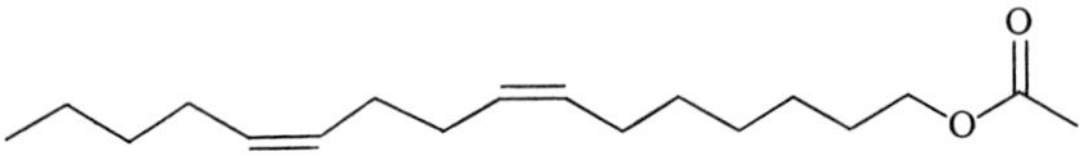

cis-7, *cis*-11-Hexadecadienyl acetate

90

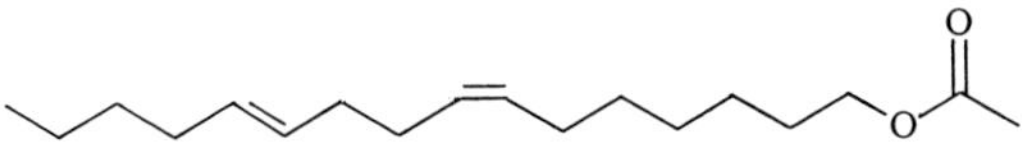

cis-7, *trans*-11-Hexadecadienyl acetate

91

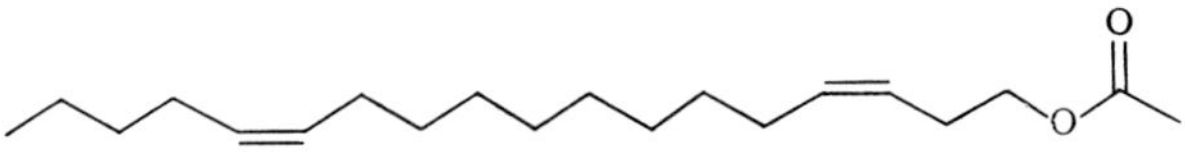

cis-3, *cis*-13-Octadecadienyl acetate

92

FIG. 28. Continued.

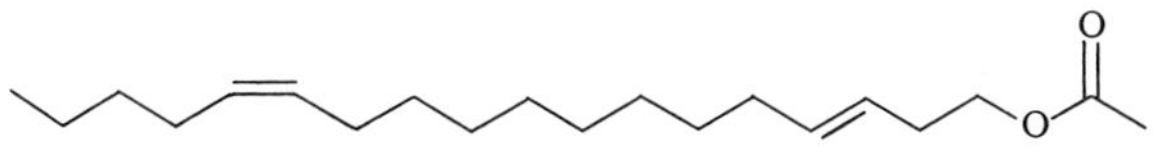

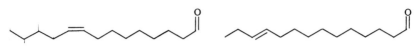

trans-3, *cis*-13-Octadecadienyl acetate

93

Aldehydes

cis-9-Tetradecenal

94

trans-11-Tetradecanal

95

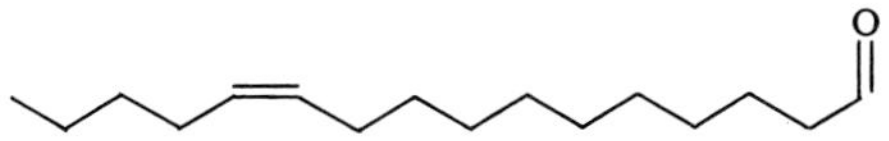

cis-11-Hexadecenal

96

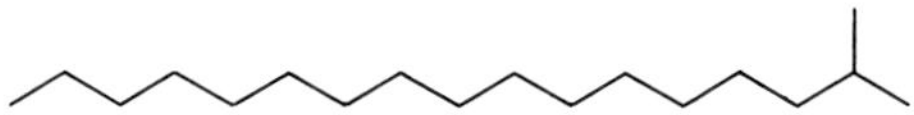

2-Methyl heptadecane

97

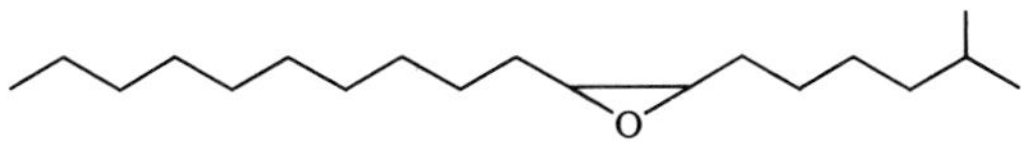

cis-7, 8-Epoxy-2-methyl octadecane

98

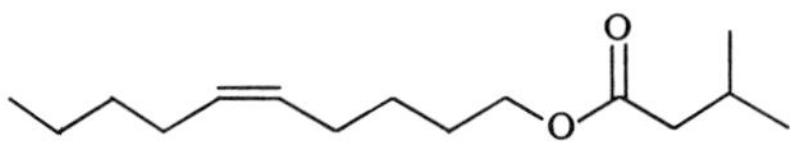

cis-5-Decenyl isovalerate

99

FIG. 28. Continued.

vestigations are conducted, many species are found to produce multiple component pheromones. Also, a few chemicals that were proposed as being pheromones were later found to be devoid of biological activity. For these reasons, the information provided in Tables 5 and 6 may require considerable modification in the future.

Most of the sex pheromone chemicals identified from female lepidopterans show a remarkable similarity in structure (Table 6). They are generally unbranched chains of 10 to 18 carbon atoms having one or two points of unsaturation and terminated by a functional group. The terminal functional group is most frequently an acetate, although it may also be an aldehyde, alcohol, or acid; it also may be lacking.

Only a few male-produced sex pheromones that cause female aggregation have been identified, corresponding to the relative rarity of species in which the male serves as the site to which the female moves prior to mating. The sex pheromone produced by male Mediterranean fruit flies consists of a mixture of methyl *trans*-6-nonenoate (**100**) and *trans*-6-nonenol (**101**) (305). That produced by males of the greater wax moth is a mixture of *n*-nonanal (**102**) and *n*-undecanal (**103**) (386, 534), and that of the lesser wax moth is a mixture of *n*-undecanal (**103**) and *n-cis*-11-octadecenal (**104**) (161).

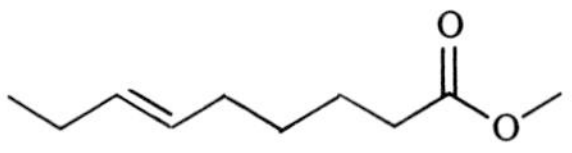

100

101

102

103

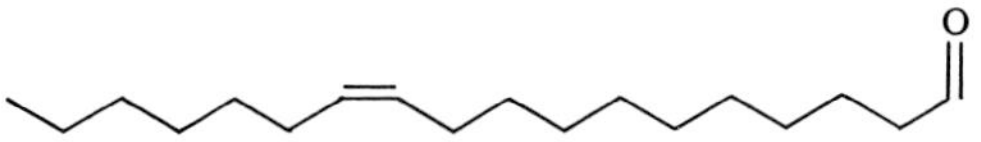

104

8.2 Stimulation of Courtship and Copulation

After the two sexes have come together, either by one sex aggregating in response to stimuli released from the other or by both sexes aggregating in the same area in response to appropriate environmental stimuli, pheromones may mediate close-range sexual behavior. Some writers refer to these pheromones as "aphrodisiacs" (64, 119). The types of close-range behavior vary greatly from species to species and include one or more of the following categories: (1) attendance of the male at the vicinity of the female until she is ready to copulate, (2) courtship behavior which enhances the likelihood that the opposite sex will copulate, (3) the assumption or maintenance by the female of the proper position for copulation, and (4) copulatory behavior by the male.

8.2.1 MALE ATTENDANCE

As mentioned earlier, female sex pheromones sometimes cause males to aggregate by arresting their locomotion after they arrive at the odor source (Chapter 4, Section 4.3.3). In some species, including certain mites and mosquitoes, the female releases the pheromone when she is still sexually immature. The stimulated males remain in attendance until the female molts to the sexual stage.

Males of some crab species perform additional behaviors that maximize the likelihood of their remaining with the females. The pheromone released from a premolt female stimulates a male to carry her under his body for long periods of time, often for a number of days. In the presence of the pheromone, the male crab extends his chelae and pulls whatever crab it comes in contact with—even another male—into the precopulatory holding position (349, 557). Kittredge *et al.* (349) have obtained evidence that the molting hormone crustecdysone (**105**) is released from the premolt female and stimulates male approach as well as this attendance behavior. The pheromone also inhibits the normal feeding responses of the males

105

(643); otherwise it is possible that no females would survive long enough to copulate.

Similar male attendance behavior occurs in the hermit crab, *Pagurus bernhardus* (267). The female pheromone is apparently perceived by the male only after he makes direct contact with her body. He then grasps the rim of her shell aperture with his minor cheliped and pulls her around with him until she is ready to copulate several days later.

8.2.2 COURTSHIP BEHAVIOR

Copulation does not automatically occur when a male and female come together, even though they might both be in the proper state of sexual readiness. Often, a further exchange of signals, referred to as courtship stimuli, is required to lower the threshold of both sexes for copulation by removing the neural inhibitions that usually keep copulatory behavior in check.

When a male lycosid spider contacts a female and perceives a sex pheromone on her cuticle, he performs an elaborate visual display, consisting of stylized leg, palpal, and antennal movements, and special ways of positioning the body. The female is thus stimulated to accept the male in copulation (180, 269).

Many male fish perform complex courtship displays when near females. For example, the display of the male gobiid fish, *Bathygobius soporator,* includes rapid fanning and gaping movements as well as a change in his body color to the courtship phase. These male behaviors and the color change are usually released by an interaction of visual and pheromonal stimuli from the female. The pheromone is a sufficiently strong stimulus that it can cause the male to perform his courtship display even when he cannot see the female (651).

Among vertebrates, the range of male courtship behaviors stimulated by female pheromones seems almost unlimited. Under the influence of the pheromone, a male western banded gecko courts by grasping the female's tail in his mouth (240). Male tortoises, *Geochelone* spp., make horizontal head movements (10). Social investigation, especially sniffing and licking of the female vaginal area, is a prominent component in mammalian courtship (14, 114, 163, 168, 170, 187, 188, 392, 445, 446). Male rhesus monkeys also spend more time grooming their sexual partner when they smell her sex pheromone (339).

8.2.3 ASSUMPTION OF THE COPULATORY POSITION BY THE FEMALE

When a receptive female pig is in the presence of a boar, she assumes a characteristic mating stance called the "immobilization reflex." A variety of stimuli from the boar, including a pheromone, sonic signals, and the pressure that he exerts on her back when mounting, cause her to assume the position (251, 604). The pheromone has a characteristic musky odor and is composed of a mixture of 16-unsaturated C_{19} steroids that are structurally related to the male hormone, testosterone (106). These chemicals are concentrated in the boar urine, saliva, and sweat (87, 232, 412, 630). Practical use has been made of two of the steroids, 5a-androst-16-en-3-one (107) and 3a-hydroxy-5a-androst-16-ene (108), which have been incorporated into an aerosol that is dispensed at the female prior to artificial insemination, causing her to automatically assume the immobilization reflex position (412).

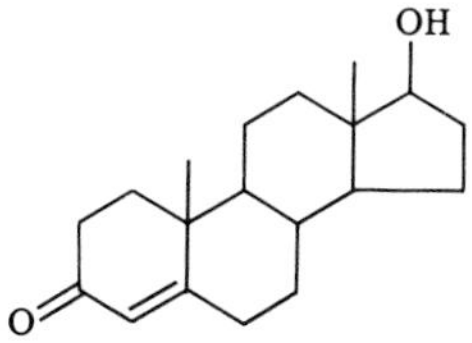

106

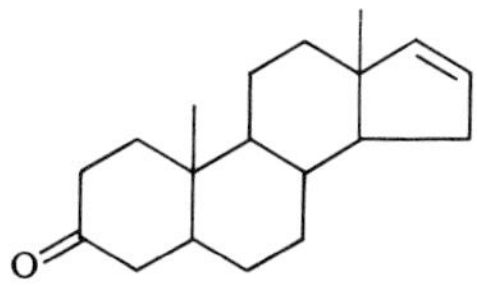

107

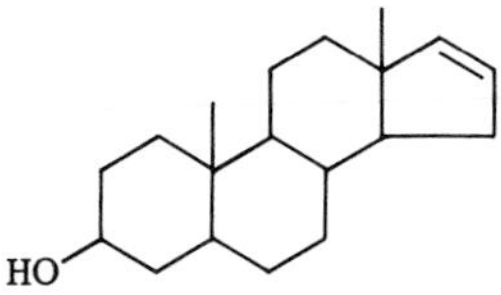

108

Males of certain millipede species present an odorous glandular area on their body to nearby females, stimulating the females to feed on the glandular secretion and in so doing assume the proper position for copulation. The male millipede pheromone also acts as an arrestant, causing a female to remain still until copulation is accomplished (246, 247). In many cockroaches, a male raises his wings when he senses a nearby female. A pheromone is then released from a gland located under his wings, causing the female to mount the male and feed on the glandular secretion. While she is in this position, the male is able to grasp her genitalia with his own and copulation ensues (30, 32, 262, 347, 540, 541, 703). A pheromone produced by male Queensland fruit flies often causes females to extrude their reproductive segments. Although this behavior has not been studied extensively, it is thought that the females might thereby be positioning themselves for copulation (199).

Unlike male moths, which usually approach distant females by sensing and following trails of sex pheromone, male butterflies typically see flying females from a distance and pursue them (25). While in flight, a male butterfly maneuvers himself into a position above the female's head so that the glands located on his wings, abdomen, or legs are positioned correctly in order to distribute pheromone over her antennae. The pheromone apparently acts as an arrestant, causing the female to alight, whereupon the male attempts to copulate with her. Often the male will disperse more pheromone toward the female's antennae prior to making copulatory attempts. Courtship in butterflies has been most intensively studied in the queen butterfly. The male's paired scent glands are located at the posterior of his abdomen. They are associated with eversible hair-like scales called scent brushes, which are fanned out during courtship and serve to distribute the scent (Fig. 29) (109, 110, 447, 489). Two components of the secretion found on the scent brushes have been identified (411). One of them, 2,3-dihydro-7-methyl-*H*-pyrrolizin-l-one (**109**), is the pheromone that acts as an arrestant of female flight activity. The pheromone is presumably adsorbed onto cuticular "dust" particles that liberally coat the scent brushes. The other component, *trans,trans*-3,7-di-

FIG. 29. Fully extruded scent brushes on the abdominal tip of a male of the queen butterfly. (Courtesy of Thomas Eisner.)

methyldeca-2,6-dien-1,10-diol (**110**), serves as a "glue" to bind the cuticular dust onto the antennae of the courted female.

The close-range courtship behavior of moths is often similar to that of butterflies. Male moths have a variety of glandular structures, depending on the species, and some of the glands are associated with eversible scent brushes (9, 25, 60–62, 64, 631). After a male moth follows the trail of female-released pheromone and arrives in her vicinity, he everts the scent brushes and then immediately attempts to copulate. Some evidence, combined with a great deal of speculation, indicates that a pheromone is released from the scales of the brushes and functions as an arrestant, preventing the female from moving away (60–62, 64, 140). Many chemicals have been identified from the scent brushes of male moths, but in most cases there is no demonstration of their biological function, and they cannot be assumed to be pheromones. However, Clearwater (140) demonstrated that benzaldehyde (**111**), a major component of the secretion produced by males of the southern armyworm moth, does indeed inhibit female movement.

| 109 | 110 | 111 |

8.2.4 MALE COPULATORY BEHAVIOR

The sterotyped motor reactions that the male exhibits during copulation may be directly stimulated by a sex pheromone released from the female. Pheromone-induced copulatory behavior occurs in such diverse groups as rotifers (222), nematodes (400), insects [partial reference lists included in Chapter 6, Section 6.3 and in (301) and (592)], tortoises (10), and mammals (168, 170, 339, 445, 446, 565).

The sex pheromone released from the vaginal area of female rhesus monkeys causes a heightening of male sexual activity, including an increase in the number of ejaculations when the males mount the females (339, 413–415). The pheromone consists of a mixture of short-chain aliphatic acids—principally acetic (**112**), propanoic (**113**), methylpropanoic (**114**), butanoic (**115**), methylbutanoic (**116**), and methylpentanoic acids (**117**) (416).

Males of some animal species do not insert sperm directly into the females. Both males and females of the polychaete worm *Platynereis dumerilii* discharge pheromones which stimulate nearby members of the opposite sex to release their genital products into the water, resulting

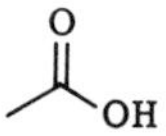

112

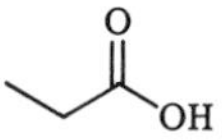

113

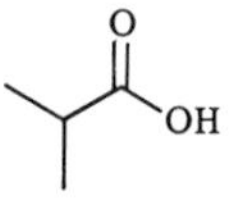

114

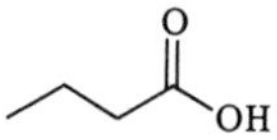

115

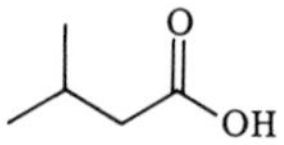

116

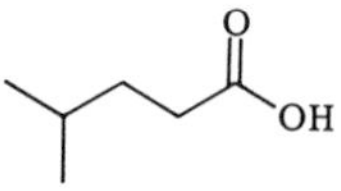

117

in external fertilization (81). Male collembolans (springtails) deposit their sperm in packets called spermatophores, which they attach by stalks to objects in the environment. Females later pick up the spermatophores and deposit them in their genital tracts. A sex pheromone produced by females of the springtail, *Sinella curviseta*, increases the deposition of spermatophores by nearby males (694).

The male copulatory reactions may be so stereotyped that they will be exhibited toward quite inappropriate objects when the female sex pheromone is perceived. Male rotifers direct copulatory movements toward pheromone-treated glass beads (222), and male houseflies attempt to copulate with pheromone-treated knots of string (Fig. 30) (533). Pheromone-stimulated males of many other insect species try to copulate with artificial female models, including 2-dimensional black silhouettes (592). A male tortoise, *Geochelone* sp., was observed attempting to mount a head of lettuce over which a female had recently clambered (10). And males of many animal species, including insects and mammals, try to copulate with nearby males when they are stimulated by the female sex pheromone (445, 446, 592).

8.3 Hierarchies of Sex Pheromone Behavior

As indicated earlier, a single sex pheromone may sometimes stimulate more than one type of behavioral reaction in an animal. The reactions often occur in a sequence or hierarchy of behavioral steps, starting with the activation of the animal, followed by its movement toward the pheromone source, and culminating with its courtship and copulatory behavior. Such hierarchies have been most often observed in male insects during their responses to female sex pheromones (11, 22, 23, 118, 142, 143, 162, 171, 196, 197, 245, 285, 293, 354, 380–383, 581, 589, 592, 598, 665, 680). For example, if a resting male of the cabbage looper moth is exposed to

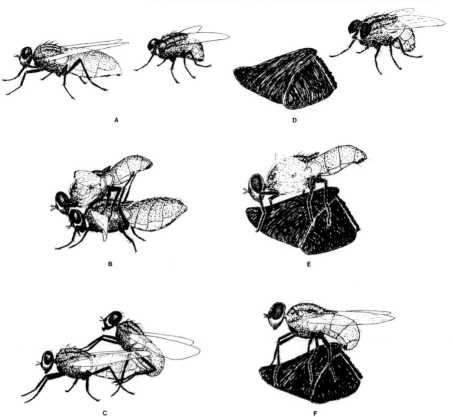

FIG. 30. Precopulatory behavior exhibited by a male housefly in response to a living female fly (A–C) or to a pseudofly made of a severed, black shoelace knot treated with female sex pheromone (D–E). The three phases of behavior shown in each case are approach (A and D), a leap onto the "mating partner," accompanied by wing vibration (B and E), and attempted copulation (C and F). [From Cowan and Rogoff (153).]

pheromone released from a female located upwind, he (1) brings his antennae forward, (2) spreads his wings, (3) vibrates his wings with increasing amplitude, (4) takes flight, (5) flies in an oriented manner toward the pheromone source, (6) reduces flight speed and hovers near the female, (7) touches the pheromone source (female's gland) with his antennae, (8) touches the pheromone source with his front feet, (9) moves to a position beside the female, everts his scent brushes, and directs them toward her, (10) attempts to copulate (Fig. 31) (230, 293, 589, 598). The female sex pheromone is often such an important stimulus of premating behavior that the remainder of the female's body is not necessary at all. If a spot of pheromone is placed on some substratum, nearby male cabbage looper moths exhibit the entire sequence of behavioral steps, including a copu-

99

FIG. 31. Males of the cabbage looper moth exhibit a sequence of behavioral steps after they perceive the sex pheromone released from a receptive female (1). When the male arrives in the vicinity of the female, he hovers below her and touches the glandular area with his antennae (2) and front feet. He then moves to a position beside her (3), everts his scent brushes and directs them toward her (4) and attempts to copulate. [From Gothilf and Shorey (230).]

latory attempt directed toward the spot from which the chemical is volatilizing.

Several factors are probably involved in determining the order in which the reactions are displayed so that the sequence correctly culminates in a male locating and copulating with a female. The behavioral steps often follow each other in a rigidly fixed succession, and it is possible that the completion of one step serves in part as a stimulus for initiation of the next (581). Also, each step may require a higher phero- mone concentration for its release than did the previous one (23, 162, 194– 197, 245, 581, 598, 665). Because some of the steps occur while the male is moving toward the female, he automatically becomes exposed to a higher pheromone concentration and thus becomes primed for the next step. The range in effective concentrations may be very large, as evidenced by the fact that a 100,000-fold increase in female sex pheromone concen- tration is needed to induce males of the light brown apple moth to attempt copulation (final step) as compared with that needed to induce antennal elevation (first step) (23). This phenomenon of many different behavioral reactions being released in sequence by a single chemical stimulus that is modulated only by different concentrations represents a marvelous economy of energy; however, other factors may also con- tribute in causing the steps to occur in the correct order. For instance, other close-range stimuli from the female probably act in conjunction with the pheromone to ensure that the final steps are displayed at the right time.

The above discussion implies that the same female-produced phero- mone operates to cause each of the behavioral steps, and indeed this often seems to be the case. However, it is also possible that in some species certain pheromone chemicals act at long range to cause male approach and other chemicals act at short range to cause courtship and copulatory reactions.

Hierarchical sequences of response to sex pheromones are probably also a common phenomenon in mammals, the males of which are often stimu- lated to approach, court, and attempt copulation with pheromone-releas- ing females. The male mammal encounters an increasing pheromone concentration as he approaches from a distance, and at close-range he may obtain a very high concentration through the common behaviors of sniffing or licking the part of the female body at which the odor originates (14, 39, 114, 168, 170, 188, 392, 445, 446, 565).

8.4 Human Sex Pheromones

In common with the other higher primates, man has a poorly devel- oped sense of smell. Until recently, monkeys and apes were considered to make little use of odorous communication. However, it now appears that in some higher primate species sex pheromones act as powerful stimulants of precopulatory behavior (186, 187, 339, 401, 413–417, 450). Similarly, although the use of sex pheromones by man is not obvious today, con- siderable evidence supports the idea that such chemicals might indeed operate as sexual stimulants, perhaps at a subliminal level, or that they operated in this manner in our ancestors.

Some of the evidence relates to morphological and physiological characteristics of man. Humans have a complete set of organs that are traditionally described as nonfunctional, but which, if seen in any other mammal, would be considered to be part of a pheromone system (146, 147). These include apocrine (skin) glands which are sometimes associated with conspicuous tufts of hairs. Some of the glands do not produce sweat and presumably have some other function. For example, the prepuce in the male and labia in the female have glands that secrete odorous materials which might be suspected as having pheromone activity.

Another bit of evidence is the fact that in many cultures humans attempt to remove all traces of natural body odors, only to replace them with perfumes that are in part based on animal secretions that may have served as sexual stimulants in the original animal species (61). These perfumes are usually supposed to make the wearer more attractive to the opposite sex, indicating that our cultural evolution might have caused us to regard our natural odors, or our sexual urges that are caused by these odors, to be repugnant and base, and to find the use of odorous materials derived from other sources to be an acceptable substitute.

Humans are particularly sensitive to compounds having the odor of musk, an odor characteristic of a number of animal secretions that have been used by men and women for anointing their bodies for at least the past 3000 years (342). Some of these compounds, including muscone, 3-methyl cyclopentadecanone (**118**), obtained from the musk deer and civetone, cis-9-cycloheptadecenone (**119**), obtained from the civet cat bear a structural resemblance to the musk-smelling compounds found in both humans and boars (**106–108**) (107, 147, 342, 369, 609).

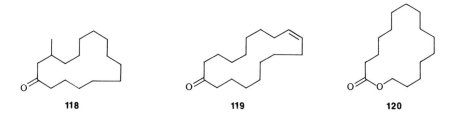

118   119   120

Women have a greater sensitivity to musk-smelling compounds than do men (359). Le Magnen (374, 375) found that the odor of the musky synthetic perfume exaltolide, 15-pentadecanolide (**120**), is perceived clearly only by sexually mature females, whose sensitivity to the compound is greatest at about the time of ovulation. Males, however, become more

sensitive to the odor of exaltolide following an injection of female hormone. This phenomenon provides strong circumstantial evidence for the
existence of an ancestral, musklike, male sex pheromone that may have
stimulated sexually receptive women.

Evidence is also accumulating to support the likelihood that human
females produce a sex pheromone in the vagina. Michael *et al.* (417)
identified acetic (**112**), propanoic (**113**), methylpropanoic (**114**), butanoic
(**115**), methylbutanoic (**116**), and methylpentanoic (**117**) acids from
vaginal samples obtained from human females. These volatile fatty acids
were the same as those shown earlier to operate as sex pheromones in
rhesus monkeys (Chapter 8, Section 8.2.4). The concentration of the
chemical varies during the menstrual cycle, reaching a peak near the
time of ovulation. Women who take oral contraceptives produce lower
amounts of the acids and do not exhibit the rhythmic changes during
the menstrual cycle.

Environmental and Physiological Control of Sex Pheromone Behavior

Most adult animals are more likely to copulate at certain times or when exposed to certain conditions than at other times or under other conditions. Likewise, the sex that produces a pheromone has a greater tendency to emit it under certain favorable conditions, and the sex that perceives the pheromone may have a greater tendency to respond when exposed to the same conditions (593). Many environmental variables interact in conjunction with the physiological state of the animal to cause sex pheromone communication to occur when conditions are appropriate for mating and to not occur when conditions are less appropriate.

9.1 Environmental Control

The most studied environmental variables—light intensity, temperature, air velocity, and surrounding vegetation—have been shown to operate directly in controlling the occurrence and timing of sex pheromone behavior. Also, some of the variables, particularly light intensity and temperature which fluctuate on a daily as well as a seasonal basis, operate indirectly by modifying the action of internal physiological clocks. The clocks then program the timing of pheromone behavior within each day or season (Chapter 9, Sections 9.2.2 and 9.2.3).

Most investigations of the direct influence of environmental factors on sex pheromone communication have been conducted on insects, with much of the emphasis placed on determining the effect of such factors on male approach responses to female-released pheromones. However, when intensive studies have been undertaken, the same factors that limit or control male responsiveness are often also found to limit or control the tendency of females to release the pheromone. Thus, the likelihood that mating will occur may be equally enhanced or inhibited in both partners in the communication sequence as a result of environmental conditions.

9.1.1 LIGHT INTENSITY

Some animals characteristically mate during the day, some during dawn or dusk, and some at night. These general times may be partially set by a light:dark-cycle-entrained physiological clock (Chapter 9, Section 9.2.3). In addition, the light intensity may have to be at an appropriate level; levels that are too high or too low may inhibit or prevent sex pheromone communication. For example, in cabbage looper moths, which mate at night, light intensities higher than that provided by full moonlight (about 0.3 lux) inhibit both the tendency of females to release pheromone (618) and that of males to respond to the pheromone (595). On the other hand, males of the light brown apple moth, which mate at dusk, remain highly responsive to female sex pheromone at light intensities as high as 3.5 lux (22). Little is known about the influence of the changing light intensities during the crepuscular periods on sex pheromone communication; however, Cardé and Roelofs (127) found that the decreasing light intensity that occurs during dusk directly stimulates females of the arctiid moth, *Holomelina immaculata,* to start releasing pheromone.

9.1.2 TEMPERATURE

Temperature may impose both upper and lower limits on the responsiveness of male moths to female sex pheromones (3, 35–37, 144, 328, 542, 590). Presumably, the tendency of females to release pheromone is restricted to the same temperature zone as that which is appropriate for male response (620). The favorable temperature zones vary according to the way of life of the species, being generally higher for day-active than for night-active forms.

Changing temperatures may also be important. A decreasing temperature interacts with the decreasing light intensity to stimulate pheromone release by females of *Holomelina immaculata* at dusk (127).

9.1.3 AIR VELOCITY

When the sex pheromone communication sequence involves flying insects, the speed of the prevailing wind often affects the tendency of one sex to release the pheromone or of the other sex to respond.

Although improper air velocities frequently prevent the responder from approaching a pheromone source, few data are available to indicate whether the lack of response is attributable to a direct inhibition of locomotory behavior when the responder senses the improper velocities, to a physical constraint on approach because velocities that are too high exceed the flight speed and force the responder back, or to a disruptive effect because velocities too high or too low make aerial trails poorly defined and difficult to follow. However, many observations indicate that the ability of male insects to approach female sex pheromone sources does become greatly impaired at very low air velocities (120, 132, 194, 328, 617), and air velocities higher than an insect's flight speed obviously prevent its upwind approach to a pheromone source (318, 622).

Presumably, within the upper and lower limiting air velocities, certain velocities are most suitable for pheromone communication, and the pheromone behavior of most animals might be primed to operate most efficiently at those velocities. Thus, females of the cabbage looper moth are more likely to release pheromones at velocities of 0.3 to 1.0 m/second (probably the optimal range for producing a stable aerial trail and allowing male upwind flight) than at higher or lower velocities (318). When they do release pheromone at lower velocities, the duration of each period of continuous release is increased, compensating for the slowness at which an aerial trail can extend downwind at those velocities.

9.1.4 SURROUNDING VEGETATION

Some animal species are relatively monophagous, feeding on only one or a few closely related plant species. Other animals have a great variety of plants on which they can feed. Monophagous species sometimes are also restricted to their host plants during pheromone communication, even though the adult animals are usually mobile and could move to other habitats (4, 495, 586, 653). This restriction may result from one or more of the following factors: (1) the females might not release pheromone unless they sense appropriate chemicals or other stimuli that indicate that they are located on the correct host (509–511); (2) the males might not respond to the pheromone unless they similarly sense that they are in the right habitat (95); (3) the males and females might only occur in the correct areas anyway, having developed there when young or being attracted there by plant-produced stimuli before the time of pheromone communication. Whatever the reason for the phenomenon, the behavior appears to be highly adaptive, ensuring that mating takes place in a plant habitat that can be utilized as food by the ensuing progeny.

The vegetation may also influence the vertical location in the habitat at which pheromone communication and mating occur. Females of many moth species cling to vegetative surfaces while releasing pheromone, and although the location of pheromone-releasing female moths has rarely been observed in nature, they are presumably positioned most often near the top of the foliage canopy. This assumption as to female location is based mainly on the fact that male moths typically respond best to synthetic pheromone sources that are located near the top of the canopy, regardless of whether the vegetation consists of cabbages or forest trees (4, 319, 419, 558, 586). The optimal vertical location of pheromone communication may, however, be influenced by air velocity. On calm nights, male pink bollworm moths are most readily attracted to female sex pheromone sources that are placed among the top leaves of cotton plants, and most mating pairs are found in this vegetative zone. Apparently, the moths can sense air speed, and on windy nights they move down in the foliage to a location at which the velocity is reduced to a level more suitable for pheromone communication (319).

9.2 Physiological Control

A number of physiological systems operating within an animal control its readiness for mating. These systems usually interact with the environmental variables discussed above to determine whether, when, and where sex pheromone communication and mating will occur.

The physiological variables include the age of the animal, internal clocks that indicate the time of day or season, influences of prior exposure to the pheromone or of prior mating, and hormones.

9.2.1 AGE

The age at which an animal produces a sex pheromone and releases it into the environment must be linked in most species to the age at which sexual maturity is attained (259, 600). The various physiological systems that regulate precopulatory behavior, copulation, and fertilization of the eggs are usually synchronized; therefore, in most animal species, juveniles do not produce sex pheromones. An exception occurs in certain female arthropods which precociously release sex pheromones before molting to the adult stage and thus cause males to remain in attendance until the final molt has been accomplished (Chapter 8, Section 8.2.1).

The same general phenomenon also applies to the responding sex. Although juveniles may in some cases have fully operative olfactory systems which allow them to perceive the pheromone (474), overt behavioral responsiveness is usually inhibited until the animal attains the physiologically mature age at which the systems that regulate precopulatory behavior, copulation, and fertilization can be synchronized (200, 259, 601). However, subadults of some mammalian species exhibit overt, but incomplete, sexual behavior in response to the pheromone released from sexually mature females. Precocious sexual behavior in squirrel monkeys, *Saimiri* sp., may enable subadult males to learn to discriminate visually between the appearance of receptive and nonreceptive females (14).

9.2.2 SEASONAL RHYTHMS

Many vertebrates and invertebrates have specific times during the year when they are reproductively active. This rhythmic behavior is well known in mammals, with the females often coming into estrus once each year and mating only during that defined period. A similar situation is seen in a wide variety of invertebrates, including planaria and crustaceans (326, 327, 506), in which the females presumably only release sex pheromones during the reproductive season. Whether sexually mature males would respond to the pheromones if perceived at inappropriate times of year is unknown.

9.2.3 DAILY RHYTHMS

Most of the behaviors of animals are exhibited only during specific time periods of each day. The daily rhythms of behavior are often controlled by endogenous timing mechanisms that are kept in phase with the time of day in the outside world through the sensing by the animal of daily changes in environmental cues such as light and darkness and that would reoccur on an approximately 24-hour periodicity even if an animal were held under constant conditions. Such endogenous rhythms are called circadian (circa = about, dian = day) rhythms.

Mating behavior is one of the activities that is frequently restricted to a certain time of day. Thus, it seems reasonable that sex pheromone release by one sex and responsiveness to the pheromone by the other should also be restricted to that same time of day because these are the processes that often initiate the mating sequence. Circadian rhythms of pheromone communication have received little or no study in vertebrates. This rhythmicity has been extensively studied in invertebrates, especially in moths in which the females generally release the sex pheromone (127, 197, 220, 340, 470, 561, 563, 618, 620, 667, 677) and the males respond (197, 225, 470, 596, 677) at a characteristic time that is controlled by a circadian rhythm. The length of time during each day when the males respond may completely overlap the length of the period during which the females are likely to release pheromone, maximizing the likelihood that the two sexes will be brought together for mating (197, 620).

Although the environmental light:dark cycle appears to be the main factor entraining the circadian rhythms of moths, temperature may determine the exact time within each 24-hour period when the rhythmic behavior is expressed. The females of a number of nocturnal species release pheromone and the males respond maximally to the pheromone near the middle of the night when the temperature is warm. Under cooler conditions, these behaviors are exhibited earlier in the dark period (127, 558, 618, 620, 669). This temperature effect may be adaptive in that communication for mating thereby occurs in the early part of the night before the temperature drops so low that activity becomes inhibited.

9.2.4 PRIOR EXPOSURE TO PHEROMONE OR PRIOR MATING

An important aspect of the behavior of most animals is that it may be modified following certain experiences. Thus, the successful culmination of sex pheromone communication (i.e., copulation) can permanently or temporarily reduce the likelihood of renewed premating communication behavior.

Females of some insect species mate only once during their adult lives. In these species, little or no additional sex pheromone may be produced or released following a single successful mating (15, 24, 144, 627, 703). In other species, the females mate more than once, and the pheromone continues to be produced after one mating (101, 597, 600), being released into the air at the beginning of the communication sequence leading to the next mating.

The pheromone responders may have a refractory period following successful mating, during which renewed exposure to the pheromone will not lead to the appropriate response. Although males of the cabbage looper moth can mate with a number of females on consecutive nights, they are limited to a single mating on any one night. Correspondingly, the males are totally nonresponsive to female sex pheromone during the remainder of a night on which they have mated (595).

Even if successful mating does not occur when an animal is exposed to and responds to a sex pheromone signal, that individual is likely to be less responsive to renewed stimulation by the odor for some time, partly as a result of an interaction of sensory adaptation and habituation to the odor. Olfactory sensory adaptation is a very transitory phenomenon wherein, for a number of seconds following a prior stimulation with the same chemicals, the sensory neurons have an increased threshold for reaction to the odor and thus a reduced tendency to relay the message to the central nervous system. The sensory neurons typically recover from adaptation within a minute after the odor stimulus has been removed (80, 474, 570).

Habituation is a central nervous system phenomenon that causes the animal to be less responsive to a pheromone stimulus for many minutes or hours following a previous response, even though the previous response did not result in mating (19, 20, 22, 46, 293, 364, 595, 621, 666, 678, 701, 702). Habituation is generally considered to be advantageous in that it prevents continued responses to a stimulus that did not lead to an appropriate reward, such as mating.

In a number of mammalian species, a male that has reached "exhaustion" after repeated copulations with a female can be aroused by and will copulate with a different female that is brought into his presence (38, 55, 112, 251, 286, 565, 714). It seems likely that the exhaustion may largely reflect habituation to the familiar stimuli, including the pheromones emanating from the original partner, and that novel stimuli from the new partner may reactivate the male's responsiveness.

A similar situation has been observed in several *Drosophila* species in which females that are courted by a number of males tend to mate with the one that is of a different kind or strain from the others (183). Since such mate selection is based partly on perception by the females of pheromones released from the males, possibly the females become habituated to the prevailing odor that is common to the majority of nearby males while remaining fully responsive to the different odor produced by the males that are rare in the population.

9.2.5 HORMONES

The hormonal and nervous systems of animals act together to coordinate the diverse reproductive processes and to ensure that the pheromone-release or pheromone-responsiveness systems are ready to function when mature gametes are available in the reproductive tract. The internal communication systems also must coordinate the timing of activity of the pheromone systems with fluctuations of the limiting environmental and physiological variables discussed above.

Among many mammals that have cyclic estrus periods, the sex hormones synchronize pheromone behavior with reproductive maturity (129-131, 163, 315, 339, 413-415). When the female is the pheromone-releasing partner, removal of the ovaries may eliminate her pheromone production and injection with estrogen may restore it. Both sexes of mice produce sex pheromones that lead to aproach of the opposite sex. In the males, the production of pheromone is probably controlled by the titer of androgen, as evidenced by the fact that only dominant males—which secrete the highest amounts of androgen—are able to attract females (314).

Females of some insect species display cyclic periods of sexual receptivity followed by nonreceptive periods of "pregnancy," similar to the reproductive cycles of mammals. For example, cyclic changes in the amount of hormone secreted from the corpora allata of cockroaches control a number of aspects of reproductive behavior, including pheromone production during the period of female receptivity (26–31, 33, 43). Females of other insect species emerge from the pupal skin, mate, deposit their eggs, and die in a short time, with no apparent hormonal control of sex pheromone production; however, even in these short-lived insects, pheromone-*release* behavior may be under such control (512). A hormone released by the corpora cardiaca when females of certain saturniid moths sense that they are located on the correct host plant or are at the right time in their circadian rhythm stimulates the ganglion in the abdominal nervous system that causes protrusion of the genitalia and the subsequent release of sex pheromone.

Hormonal control of sex pheromone production reaches its peak in *Pachygrapsus* and *Cancer* crabs, with the hormone also serving as the pheromone. Some of the crustecdysone [(105), Chapter 8, Section 8.2.1] that is involved in the process of molting to the adult female stage is released with the urine into the water and stimulates males to approach and to remain in attendance until molting has been completed (349).

10

Sex Pheromones and Reproductive Isolation

Reproductive isolation mechanisms can operate at ecological, behavioral, physiological, and/or morphological levels to prevent fertile matings between males of one species and females of another. Examples of how some of these mechanisms operate include the following:

1. The potential cross-breeding species may be isolated from each other because they occur in different geographic areas or in different habitats, or, if they occur in the same area, because they are sexually active at different seasons or different times of day.

2. The different species may have independent long-range communication systems so that potential responders of one species do not approach a signal emitter of another.

3. Differences in short-range courtship behavior between species may prevent cross mating, even if the two sexes do come together.

4. Incompatibilities may occur at the morphological level, so that successful interspecific copulation cannot physically occur, or at the physiological level, so that even though a male of one species successfully copulates with a female of another, the interspecific hybrids are nonviable.

Examples 1 and 2, the most efficient mechanisms since they prevent inappropriate partners from ever coming together and thus avoid wastage of energy and gametes, have been extensively studied with respect to the sex-pheromone communication systems of moths.

Time-of-day isolation enables more than one species to communicate over a distance using the same pheromone chemicals and yet avoid cross attraction. The characteristic time of pheromone release by females and of pheromone responsiveness by males may be compartmentalized into a few hours of the night that are different from the hours used by related species (145, 231, 321, 338, 498). For example, although females of both the alfalfa looper moth and the cabbage looper moth apparently release

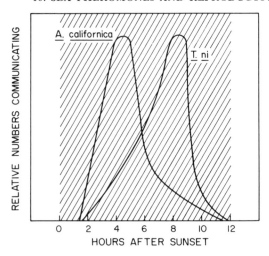

FIG. 32. Times of initiation of pheromone communication by females of the alfalfa looper moth, *Autographa californica*, and the cabbage looper moth, *Trichoplusia ni*. The shaded area indicates the time between sunset and dawn. [Adapted from Kaae *et al.* (321).]

the same chemical as a sex pheromone, reproductive isolation may be partially brought about because of a tendency toward different times during the night when the pheromone communication of these two species occurs (Fig. 32).

When related moth species occur in the same habitat and mate during the same season and time of day, their long-range pheromone communication systems are typically so distinct and specific that males of one species are not attracted to females of another. However, this is not an absolute rule, and there are a number of examples of cross-species pheromone attraction (32, 49, 211, 461, 491, 562, 580, 599) in which close-range courtship isolation or incompatibilities at the morphological or physiological level must operate to prevent successful mating from occurring.

The most obvious factor causing long-range pheromone communication systems of related species to be distinct is the use of one or more chemicals by one species that are different from the chemicals used by related species. In some cases, extreme specificity occurs, with the sensory systems of male moths being so finely tuned that even the most minute modification in the chemical structure of the pheromone of that species causes a drastic, if not complete, loss in the ability of the male to perceive the chemical (2, 49, 217, 239, 303, 564, 567, 569, 572, 639, 662). Thus, although females of related species might release pheromones having very closely related structures, there is likely to be no cross attraction because the males might not even perceive the presence of pheromones belonging to the wrong species. In other cases, the males of one species do perceive the closely related pheromone chemicals released from the wrong species, but these related chemicals act as negative stimuli, inhibiting the male's responses. Such a cross-inhibition effect appears to maintain reproductive isolation between the lesser peach tree borer moth and the related peach tree borer moth (674). The females of these species release different geometric isomers of 3,13-octadecadienyl acetate [Chapter 8, Section 8.1 (Fig. 28, **92** and **93**)]. The "correct" isomer—that isomer which is released by the females of one of the species—is attractive to males of the same species when evaporated from synthetic sources into the air. The incorrect isomer, produced by females of the wrong species, is so inhibitory that it prevents males from even approaching pheromone-releasing females of their own species when evaporated into the surrounding air.

When more than one chemical occurs in the pheromone of each species, the sophistication of the isolating mechanisms can be increased. Some groups of related moth species "share" one chemical component, which is attractive to all of the males in a group if it is released into the air by itself; however, other chemicals, which are often present in the females in relatively small quantities, enhance the attractiveness of the pheromone for males of the correct species and at the same time inhibit attraction of males of the wrong species (21, 145, 210, 422, 522–524, 623, 624, 642, 645, 673).

Sometimes the related species use some or all of the same multiple pheromone components; in these cases, a characteristic quantitative ratio of the components for each species is the factor that may ensure species specificity (357, 420, 421, 645–649). The separate chemicals in the pheromone blend are often nonattractive if released alone, and the correct ratio of components is so specific that the ratio used by a related species may also be essentially nonattractive. Roelofs et al. (530) listed the various moth species (see tabulation below) that are known to use multicomponent pheromones consisting of the cis and trans isomers of 11-tetradecenyl acetate (Fig. 28, **86** and **87**) and the ratios of these two chemicals that are most attractive to the males of each of the species.

cis : trans	Moth species	Reference
97:3	European cornborer (Iowa population)	(357)
92:8	Red-banded leaf roller	(357, 532)
70:3	Fruit tree leaf roller	(530)
50:50	Fruit tree tortrix	(480)
50:50	Smartweed borer	(357)
10:90	Omnivorous leaf roller	(278)
3:97	European cornborer (New York population)	(358)

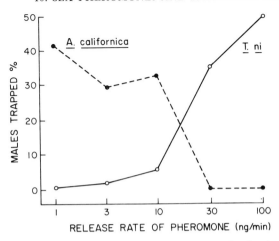

FIG. 33. Pheromone concentration as a mechanism for reproductive isolation between two lepidopterous species, the alfalfa looper moth, *Autographa californica,* and the cabbage looper moth, *Trichoplusia ni.* Percentages of males of the two species entering traps baited with their common sex pheromone differ according to the concentrations of pheromone leaving the traps. [Adapted from Kaae, R. S., Shorey, H. H., and Gaston, L. K. (1973). Pheromone concentration as a mechanism for reproductive isolation between two lepidopterous species. *Science* **179,** 487–488. Copyright by the American Association for the Advancement of Science, 1973.]

Finally, critical concentration levels of shared pheromones may minimize attraction between related species (320, 356). Both the cabbage looper moth and the alfalfa looper moth apparently use *cis*-7-dodecenyl acetate (Fig. 28, **78**) as their sex pheromone. Alfalfa looper females contain a much lower quantity of the chemical in their glands and probably release a correspondingly lower quantity into the air than do cabbage looper females; thus, it follows that low release rates attract mainly alfalfa looper males and higher release rates attract mainly cabbage looper males (Fig. 33) (320), although the mechanisms preventing alfalfa looper males from approaching sources releasing high quantities of the pheromone are unknown.

11

Evolution of Pheromonal Communication

Pheromone communication is a primitive biological function that probably arose at a very early stage during the evolution of single-celled organisms (253, 710).We might speculate that such organisms first developed a responsiveness to chemicals as a means for locating the other organisms that served as their food. With the advent of sexuality, the organisms were then preadapted for intraspecific chemical communication, and the strong selective pressure to come together for the exchange of genetic material led to a rapid specialization of these early sex pheromone systems. Even today, chemical communication is probably the major mechanism allowing single-celled plants and animals as well as the motile gametes of higher animals to orient toward a sexual partner (198, 439).

The primordial chemical communication between free-living cells was probably a necessary precursor to the evolution of metazoans. After all, the cells of metazoans might be regarded as organisms (not free-living) that exist together with division of labor in a type of society, the animal body, and that communicate with each other primarily by chemical transmitters and hormones. In this light, the chemical communication between the cells that make up a complex organism may have evolved from chemical communication systems that were used between individual free-living cells (252, 709). An example of an intermediate stage might be seen in the cellular slime molds (Chapter 5, Section 5.6), in which pheromones released by the individual cells cause them to aggregate together and form a simple multicellular organism.

Although the pheromones of primitive, free-living cells may have become internalized as the chemical messengers that are used between the cells of metazoans, an obvious strong selection pressure remained for many metazoans to continue to have a means for external, pheromonal communication with others of their species. In the metazoans, specialized groups of cells, usually comprising epidermal glands, have taken over the role of producing and releasing the pheromones to the exterior. Judging by the proliferation of pheromone systems in certain insect and mammalian species, this pressure must have been particularly strong in the more social animals.

Wynne-Edwards (724) has suggested that all mammalian pheromones were derived by natural selection from metabolites that were originally produced for some other function. For instance, the epidermal glands that originally supplied wax or mucus to the skin have probably often been later modified to produce pheromones.

Many of the chemicals that serve as alarm pheromones and cause appropriate survival responses in others of the species have a dual function (Chapter 5, Section 5.3 and Chapter 7, Section 7.1), often also serving as defensive compounds that are toxic or repugnant to the potential enemy whose presence led to their release. We might speculate that the defensive role was the first function of such chemicals and that later selection led to the use of these and related chemicals for exciting alarm behavior in conspecifics (221).

The link between external and internal chemical communication remains apparent in many complex, present-day animals, but now it appears that in many cases hormonal systems have served as evolutionary precursors of pheromonal systems. Terpenoid and/or steroid compounds are employed as either hormones or pheromones by many vertebrate and invertebrate species. Sometimes the same compound exerts overlapping hormonal and pheromonal action in the same animal species (379). Thus, it appears that the identical steroid compound, crustecdysone [Chapter 8, Section 8.2.1 (**105**)], that serves as a molting hormone for certain crabs is released by premolt females into the water, where it excites precopulatory behavior among males. In this case, the hormone was probably preexisting and later became adapted for the pheromone function through the development of a mechanism for release of the chemical into the water from females at the appropriate time, and externalization of receptors for receipt of the chemical message on the antennules of males (348, 349).

The sex hormones of some animals, or their metabolites, are probably released to the exterior, where they serve as sex pheromones and cause appropriate premating responses by members of the opposite sex. Amouriq (6, 7) has indicated that the sex pheromone of a female guppy is probably the female sex hormone, estrogen, that is released into the water through her gonopore when she is in the correct physiological state for mating.

Musk-smelling androsterols are found in both humans and boars (Chapter 8, Sections 8.2.3 and 8.4). These compounds are closely related to, and may be metabolic by-products of, the male sex hormone androgen. The submaxillary glands and salivary glands of boars appear to be target organs for androgen, with the androsterols being released from the glands as pheromones during premating behavior (87). This likelihood of a direct relationship between hormonal steroids and sex pheromones in higher mammals, and possibly in man, is intriguing (600).

In some animal species, present-day pheromone communication systems may have evolved from the olfactory responses that their evolutionary precursors made to chemicals secreted by the living or nonliving material that served as their aggregation sites. For example, the odor of ripe bananas causes males and females of *Drosophila melanogaster* to aggregate for feeding or egg deposition. The odor also excites the males sexually, stimulating them to approach nearby flies—either males or females—in their normal precopulatory manner (603). A similar phenomenon occurs in the sheep blowfly; males exposed to the odor of sheep's liver exhibit copulatory behavior toward nearby flies. Shorey *et al.* (603) proposed that an evolutionary sequence may have started with the odor of suitable host material merely bringing both sexes together, thus increasing the likelihood that they would detect close-range stimuli emanating from each other and mate. Later the same host odor could have stimulated male sexual behavior, and still later the odor or its metabolites could have been released from females that had been fed upon the host material, now making sexual communication independent of the original aggregation site. Further evolution, especially driven by the strong selective pressure for avoidance of communicative interactions with closely related species or with other strains within the same species, could have led to the great diversity and specificity of the pheromones found in many groups of present-day animals (Chapter 8, Section 8.1, Tables 4 and 5) (13, 357, 367, 427, 483, 521, 691).

A striking resemblance is noted between the odors of the host trees and the pheromones of many bark beetles (Chapter 5, Section 5.5). The host odors, which are terpene resins, are often attractive in themselves, and some of the pheromones are attractive when released alone. However, for many scolytid species, medleys of the host odors and insect-produced pheromones operate together to induce aggregation. The pheromone compounds have not been found in the host trees, although they are apparently oxidative metabolites of the host terpenes (289). For example, myrcene (**121**), found in pine tree resin, appears to be a precursor of ipsdienol (**122**), a pheromone chemical produced by males of certain *Ips* species. In turn, ipsdienol may serve as a precursor for ipsenol (**123**), another chemical used in the *Ips* pheromone blend (290).

Much more knowledge concerning the chemistry of animal pheromones and their biosynthetic pathways is needed to aid our understanding of the evolutionary history of these compounds. In most cases, the

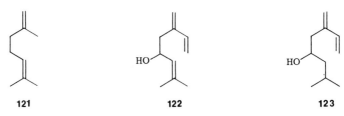

121   122   123

chemical nature of the pheromones is not known at all, although much may be known about their behavioral functions. Mice represent a good example of social mammals in which a great variety of communicative functions have been identified, although they are served by totally unidentified pheromones. Most of the functions and references have been given in the preceding chapters, but a brief listing is provided again to illustrate the variety and complexity of mouse pheromone communication. Pheromones in mice may be released by:

1. The various individuals as distinct blends of compounds, allowing recognition by others in the society
2. Stressed individuals, stimulating dispersion of other mice
3. Females, stimulating the approach and sexual behavior of males
4. Females, inhibiting the aggressive tendencies of males
5. "Foreign" females, stimulating aggression by other females
6. Lactating females, stimulating the approach of preweanling young
7. Males, stimulating the approach of females
8. "Foreign" males, stimulating aggression by other males
9. Males that coexist in a territory, inhibiting the aggressive tendencies of familiar males
10. Males that coexist in a territory, inhibiting the aggressive tendencies of foreign males
11. Males that coexist in a territory, stimulating foreign males to not enter that area

This listing is of releaser pheromones only and does not include the primer pheromones that are produced by males and regulate the reproductive physiology of females. It is unlikely that there are 11 distinct pheromones as inferred above; rather, more than one function is probably served by the same secretion. Thus, the pheromone that is released by females and that stimulates male sexual behavior may also inhibit male aggressive tendencies. Also, the pheromone that inhibits foreign males from entering a territory is probably the same as that which inhibits their aggressive tendencies. On the other hand, further study will probably reveal additional behavioral functions served by pheromone communication in mice.

One might wonder which came first, the development of social organization among animals or the proliferation of inter-individual communication systems. There is, of course, no straightforward answer to this question, because advanced levels of sociality surely applied selective pressures in favor of more sophisticated and diverse communication systems, and the development of the communication systems allowed the evolution of even more advanced societies.

Chemical communication appears to be more important for most animals than are the other communication modes. Blum (70) stated that "the road to insect sociality was paved with pheromones." Although there are many exceptions, Blum's statement should be expanded to include much of the animal kingdom.

Bibliography

1. Abdullah, M. (1965). *Protomeloe crowsoni,* a new species of a new tribe (Proto-meloini) of the blister beetles (*Coleoptera, meloidae*), with remarks on a postulated new pheromone (cantharidin). *Entomol. Tidskr.* **86,** 43–48.
2. Adler, V. E., Beroza, M., Bierl, B. A., and Sarmiento, R. (1972). Electroantenno-grams and field attraction of the gypsy moth sex attractant disparlure and related compounds. *J. Econ. Entomol.* **65,** 679–681.
3. AliNiazee, M. T., and Stafford, E. M. (1971). Evidence of a sex pheromone in the omnivorous leaf roller, *Platynota stultana* (Lepidoptera:Tortricidae): Laboratory and field testing of male attraction to virgin females. *Ann. Entomol. Soc. Am.* **64,** 1330–1335.
4. AliNiazee, M. T., and Stafford, E. M. (1972). Sex pheromone studies with the omni-vorous leaf roller, *Platynota stultana* (Lepidoptera:Tortricidae): Effect of various environmental factors on attraction of males to the traps baited with virgin females. *Ann. Entomol. Soc. Am.* **65,** 958–961.
5. Alphey, T. J. (1971). Studies on the aggregation behaviour of *Nippostrongylus brasiliensis. Parasitology* **63,** 109–117.
6. Amouriq, L. (1965). L'activité et le phénomène social chez *Lebistes reticulatus* (Poeciliidae-Cyprinoidontiformes). *Ann. Sci. Nat., Zool. Biol. Anim.* [12] **7,** 151–172.
7. Amouriq, L. (1965). Origine de la substance dynamogène émise par *Lebistes reticu-latus* femelle (Poisson *Poeciliidae,* Cyprinodontiformes). *C. R. Hebd. Seances Acad. Sci.* **260,** 2334–2335.
8. Antich, A. V. (1968). Atraccion por en ninfas y adultos de *Rhodnius prolixus* (Stal). *Rev. Inst. Med. Trop. Sao Paulo* **10,** 242–246.
9. Aplin, R. T., and Birch, M. C. (1968). Pheromones from the abdominal brushes of male noctuid Lepidoptera. *Nature (London)* **217,** 1167–1168.
10. Auffenberg, W. (1965). Sex and species discrimination in two sympatric South American tortoises. *Copeia* [N.S.] **1965,** 335–342.
11. August, C. J. (1971). The role of male and female pheromones in the mating be-haviour of *Tenebrio molitor. J. Insect Physiol.* **17,** 739–751.
12. Ayre, G. L., and Blum, M. S. (1971). Attraction and alarm of ants (*Camponotus* spp.- Hymenoptera:Formicidae) by pheromones. *Physiol. Zool.* **44,** 77–83.
13. Balakanich, S., and Samoiloff, M. R. (1974). Development of nematode behavior: Sex attraction among different strains of free-living *Panagrellus redivivius. Can. J. Zool.* **52,** 835–845.
14. Baldwin, J. D. (1970). Reproductive synchronization in squirrel monkeys (Saimiri). *Primates* **11,** 317–326.

15. Banerjee, A. C. (1969). Sex attractants in sod webworms. *J. Econ. Entomol.* **62,** 705–708.

16. Barbier, J., and Lederer, E. (1960). Structure chimique de la substance royale de la reine d'abeille (*Apis mellifera* L.) *C. R. Hebd. Seances Acad. Sci.* **250,** 4467–4469.

17. Bardach, J. E., and Todd, J. H. (1970). Chemical Communication in fish. *Adv. Chemoreception* **1,** 205–240.

18. Barrows, E. M. (1974). Aggregation behavior and response to sodium chloride in females of a solitary bee, *Augochlora pura* (Hymenoptera:Halictidae). *Fla. Entomol* **57,** 189–194.

19. Bartell, R. J., and Lawrence, L. A. (1973). Reduction in responsiveness of males of *Epiphyas postvittana* (Lepidoptera) to sex pheromone following previous brief pheromonal exposure. *J. Insect Physiol.* **19,** 845–855.

20. Bartell, R. J., and Roelofs, W. L. (1973). Inhibition of sexual response in males of the moth *Argyrotaenia velutinana* by brief exposures to synthetic pheromone or its geometrical isomer. *J. Insect Physiol.* **19,** 655–661.

21. Bartell, R. J., and Roelofs, W. L. (1973). Evidence for natural secondary compounds which modify the response of males of the redbanded leafroller to female sex pheromone. *Ann. Entomol. Soc. Am.* **66,** 481–483.

22. Bartell, R. J., and Shorey, H. H. (1969). A quantitative bioassay for the sex pheromone of *Epiphyas postvittana* (Lepidoptera) and factors limiting male responsiveness. *J. Insect Physiol.* **15,** 33–40.

23. Bartell, R. J., and Shorey, H. H. (1969). Pheromone concentrations required to elicit successive steps in the mating sequence of males of the light-brown apple moth, *Epiphyas postvittana. Ann. Entomol. Soc. Am.* **62,** 1206–1207.

24. Bartell, R. J., Shorey, H. H., and Barton Browne, L. (1969). Pheromonal stimulation of the sexual activity of males of the sheep blowfly *Lucilia cuprina* (Calliphoridae) by the female. *Anim. Behav.* **17,** 576–585.

25. Barth, R. (1958). Estimulos quimicos como meio de comunicacão entre os sexos em Lepidopteros. *An. Acad. Bras. Cienc.* **30,** 343–362.

26. Barth, R. H., Jr. (1961). Hormonal control of sex attractant production in the Cuban cockroach. *Science* **133,** 1598–1599.

27. Barth, R. H. (1962). The endocrine control of mating behavior in the cockroach *Byrsotria fumigata* (Guérin). *Gen. Comp. Endocrinol.* **2,** 53–69.

28. Barth, R. H., Jr. (1963). Endocrine-exocrine mediated behavior in insects. *Proc. Int. Congr. Zool., 16th, 1963* Vol. 3, pp. 3–5 (abstr.).

29. Barth, R. H. (1965). Insect mating behavior: Endocrine control of a chemical communication system. *Science* **149,** 882–883.

30. Barth, R. H., Jr. (1968). The comparative physiology of reproductive processes in cockroaches. Part I. Mating behaviour and its endocrine control. *Adv. Reprod. Physiol.* **3,** 167–207.

31. Barth, R. H., Jr. (1970). Pheromone-endocrine interactions in insects. *Mem. Soc. Endocrinol.* **18,** 373–404.

32. Barth, R. H. (1970). The mating behavior of *Periplaneta americana* (Linnaeus) and *Blatta orientalis* Linnaeus (Blattaria, Blattinae), with notes on 3 additional species of *Periplaneta* and interspecific action of female sex pheromones. *Z. Tierpsychol.* **27,** 722–748.

33. Barth, R. H., and Lester, L. J. (1973). Neuro-hormonal control of sexual behavior in insects. *Annu. Rev. Entomol.* **18,** 445–472.

34. Barton Browne, L., Bartell, R. J., and Shorey, H. H. (1969). Pheromone-mediated behavior leading to group oviposition in the blowfly *Lucilia cuprina. J. Insect Physiol.* **15,** 1003–1014.

35. Batiste, W. C. (1970.) A timing sex-pheromone trap with special reference to codling moth collections. *J. Econ. Entomol.* **63**, 915–918.
36. Batiste, W. C., Olson, W. H., and Berlowitz, A. (1973). Codling moth: Influence of temperature and daylight intensity on periodicity of daily flight in the field. *J. Econ. Entomol.* **66**, 883–892.
37. Batiste, W. C., Olson, W. H., and Berlowitz, A. (1973). Codling moth: Diel periodicity of catch in synthetic sex attractant vs. female-baited traps. *Environ. Entomol.* **2**, 673–676.
38. Beamer, W., Bermant, G., and Clegg, M. T. (1969). Copulatory behaviour of the ram, *Ovis aries.* II. Factors affecting copulatory satiation. *Anim. Behav.* **17**, 706–711.
39. Beauchamp, G. K. (1973). Attraction of male guinea pigs to conspecific urine. *Physiol. Behav.* **10**, 589–594.
40. Beauchamp, G. K., and Beruter, J. (1973). Source and stability of attractive components in guinea pig (*Cavia porcellus*) urine. *Behav. Biol.* **9**, 43–47.
41. Beavers, J. B., and Hampton, R. B. (1971). Growth, development, and mating behavior of the citrus red mite (Acarina:Tetranychidae). *Ann. Entomol. Soc. Am.* **64**, 804–806.
42. Bédard, W. D., Tilden, P. E., Wood, D. L., Silverstein, R. M., Brownlee, R. G., and Rodin, J. O. (1969). Western pine beetle: Field response to its sex pheromone and a synergistic host terpene, myrcene. *Science* **164**, 1284–1285.
43. Bell, W. J., and Barth, R. H., Jr. (1970). Quantitative effects of juvenile hormone on reproduction in the cockroach *Byrsotria fumigata. J. Insect Physiol.* **16**, 2303–2313.
44. Bell, W. J., Parsons, C., and Martinko, E. A. (1972). Cockroach aggregation pheromones: Analysis of aggregation tendency and species specificity (Orthoptera:Blattidae). *J. Kans. Entomol. Soc.* **45**, 414–421.
45. Bell, W. J., Burk, T., and Sams, G. R. (1973). Cockroach aggregation pheromone: Directional orientation. *Behav. Biol.* **9**, 251–255.
46. Bell, W. J., Burns, R. E., and Barth, R. H. (1974). Quantitative aspects of the male courting responses in the cockroach *Byrsotria fumigata* (Guérin) (Blattaria). *Behav. Biol.* **10**, 419–433.
47. Bennett, R. B., and Borden, J. H. (1971). Flight arrestment of tethered *Dendroctonus pseudotsugae* and *Trypodendron lineatum* (Coleoptera:Scolytidae) in response to olfactory stimuli. *Ann. Entomol. Soc. Am.* **64**, 1273–1286.
48. Berger, R. S. (1966). Isolation, identification, and synthesis of the sex attractant of the cabbage looper, *Trichoplucia ni. Ann. Entomol. Soc. Am.* **59**, 767–771.
49. Berger, R. S., and Canerday, T. D. (1968). Specificity of the cabbage looper sex attractant. *J. Econ. Entomol.* **61**, 452–454.
50. Bergström, G., and Löfqvist, J. (1972). Similarities between the Dufour's gland secretions of the ants *Camponotus ligniperda* (Latr.) and *Camponotus herculeanus* (L.). *Entomol. Scand.* **3**, 225–238.
51. Bergström, G., and Löfqvist, J. (1972). *Camponotus ligniperda* Latr.—a model for the composite volatile secretions of Dufour's gland in Formicine ants. *In* "Chemical Releasers in Insects" (A. S. Tahori, ed.), pp. 195–223. Gordon & Breach, New York.
52. Bergström, G., and Löfqvist, J. (1973). Chemical congruence of the complex odoriferous secretions from Dufour's gland in three species of ants of the genus *Formica. J. Insect Physiol.* **19**, 877–907.
53. Bergström, G., Kullenberg, B., Ställberg-Stenhagen, S., and Stenhagen, S. (1968). Studies on natural odoriferous compounds. II. Identification of a 2,3-dihydrofarnesol as the main component of the marking perfume of male bumble bees of the species *Bombus terrestris* L. *Ark. Kemi* **28**, 453–469.

54. Berisford, C. W., and Brady, U. E. (1972). Attraction of Nantucket pine tip moth males to the female sex pheromone. *J. Econ. Entomol.* **65**, 430–433.
55. Bermant, C., Clegg, M. T., and Beamer, W. (1969). Copulatory behaviour of the ram, *Ovis aries*. I. A normative study. *Anim. Behav.* **17**, 700–705.
56. Beroza, M., Bierl, B. A., and Moffit, H. R. (1974). Sex pheromones: (E,E)-8,10-Dodecadien-1-ol in the codling moth. *Science* **183**, 89–90.
57. Bethe, von A. (1932). Vernachlässigte Hormone. *Naturwissenschaften* **20**, 177–181.
58. Bierl, B. A., Beroza, M., and Collier, C. W. (1972). Isolation, identification, and synthesis of the gypsy moth sex attractant. *J. Econ. Entomol.* **65**, 659–664.
59. Birch, A, J., Brown, W. V., Corrie, J. E. T., and Moore, B. P. (1972). Neocembrene-A, a termite trail pheromone. *J. Chem. Soc., Perkin Trans. 1.* **21**, 2653–2658.
60. Birch, M. C. (1970). Structure and function of the pheromone-producing brush-organs in males of *Phlogophora meticulosa* (L.) (Lepidoptera:Noctuidae). *Trans. R. Entomol. Soc. London* **122**, 277–292.
61. Birch, M. C. (1970). Persuasive scents in moth sex life. *J. Am. Mus. Nat. Hist.* **79**, 34–39, cont. p. 72.
62. Birch, M. C. (1970). Pre-courtship use of abdominal brushes by the nocturnal moth, *Phlogophora meticulosa* (L.) (Lepidoptera:Noctuidae). *Anim. Behav.* **18**, 310–316.
63. Birch, M. C. (1974). Introduction. *In* "Pheromones" (M. C. Birch, ed.), pp. 1–7. Am. Elsevier, New York.
64. Birch, M. C. (1974). Aphrodisiac pheromones in insects. *In* "Pheromones" (M. C. Birch, ed.), pp. 115–134. Am. Elsevier, New York.
65. Birch, M. C., ed. (1974). "Pheromones." Am. Elsevier, New York.
66. Birch, M., Trammel, K., Shorey, H. H., Gaston, L. K., Hardee, D. D., Cameron, E. A., Sanders, C. J., Bédard, W. D., Wood, D. L., Burkholder, W. E., and Müller-Schwarze, D. (1974). Programs utilizing pheromones in survey or control. *In* "Pheromones" (M. C. Birch, ed.), pp. 411–461. Am. Elsevier, New York.
67. Blackford Cook, S. (1969). Experiments on homing in the limpet *Siphonaria normalis*. *Anim. Behav.* **17**, 679–682.
68. Blum, M. S. (1966). Chemical releasers of social behavior. VIII. Citral in the mandibular gland secretion of *Lestrimelitta limao* (Hymenoptera:Apoidea:Melittidae). *Ann. Entomol. Soc. Am.* **59**, 962–964.
69. Blum, M. S. (1969). Alarm pheromones. *Annu. Rev. Entomol.* **14**, 57–80.
70. Blum, M. S. (1974). Pheromonal bases of social manifestations in insects. *In* "Pheromones" (M. C. Birch, ed.), pp. 190–199. Am. Elsevier, New York.
71. Blum, M. S. (1974). Pheromonal sociality in the Hymenoptera. *In* "Pheromones" (M. C. Birch, ed.), pp. 222–249. Am. Elsevier, New York.
72. Blum, M. S., and Brand, J. M. (1972). Social insect pheromones: Their chemistry and function. *Am. Zool.* **12**, 553–576.
73. Blum, M. S., and Portocarrero, C. A. (1964). Chemical releasers of social behavior. IV. The hindgut as the source of the odor trail pheromone in the neotropical army ant genus *Eciton*. *Ann. Entomol. Soc. Am.* **57**, 793–794.
74. Blum, M. S., and Warter, S. L. (1966). Chemical releasers of social behavior. VII. The isolation of 2-heptanone from *Conomyrma pyramica* (Hymenoptera:Formididae:Dolichoderinae) and its modus operandi as a releaser of alarm and digging behavior. *Ann. Entomol. Soc. Am.* **59**, 774–779.
75. Blum, M. S., Crewe, R. M., Kerr, W. E., Keith, L. H., Garrison, A. W., and Walker, M. M. (1970). Citral in stingless bees: Isolation and functions in trail-laying and robbing. *J. Insect Physiol.* **16**, 1637–1648.
76. Boch, R., and Morse, R. A. (1974). Discrimination of familiar and foreign queens by honey bee swarms. *Ann. Entomol. Soc. Am.* **67**, 709–711.

77. Boch, R., and Shearer, D. A. (1962). Identification of geraniol as the active component in the Nassanoff pheromone of the honey bee. *Nature (London)* **194,** 704–706.

78. Boch, R., and Shearer, D. A. (1965). Alarm in the beehive. *Am. Bee J.* **105,** 206–207.

79. Boch, R., and Shearer, D. A., (1965). Attracting honey bees to crops which require pollination. *Am. Bee J.* **105,** 166–167.

80. Boeckh, J., Kaissling, K. E., and Schneider, D. (1965). Insect olfactory receptors. *Cold Spring Harbor Symp. Quant. Biol.* **30,** 263–280.

81. Boilly-Marer, Y. (1974). Etude expérimentale du comportement nuptial de *Platynereis dumerilii* (Annelida:Polychaeta): Chémoréception, émission des produits génitaux. *Mar. Biol.* **24,** 167–179

82. Bonner, J. T. (1963). How slime molds communicate. *Sci. Am.* **209,** 84–93.

83. Bonner, J. T. (1970). The chemical ecology of cells in the soil. *In* "Chemical Ecology" (E. Sondheimer and J. B. Simeone, eds.), pp. 1–19. Academic Press, New York.

84. Bonner, J. T., and Hoffman, M. E. (1963). Evidence for a substance responsible for the spacing pattern of aggregation and fruiting in the cellular slime molds. *J. Embryol. Exp. Morphol.* **11,** 571–589.

85. Bonner, T. P., and Etges, F. J. (1967). Chemically mediated sexual attraction in *Trichinella spiralis. Exp. Parasitol.* **21,** 53–60.

86. Booth, D. C., and Lanier, G. N. (1974). Evidence of an aggregating pheromone in *Pissodes approximatus* and *P. strobi. Ann. Entomol. Soc. Am.* **67,** 992–994.

87. Booth, W. D., Hay, M. F., and Dott, H. M. (1973). Sexual dimorphism in the submaxillary gland of the pig. *J. Reprod. Fertil.* **33,** 163–166.

88. Borden, J. H. (1967). Factors influencing the response of *Ips confusus* (Coleoptera: Scolytidae) to male attractant. *Can. Entomol.* **99,** 1164–1193.

89. Borden, J. H. (1974). Aggregation pheromones in the Scolytidae. *In* "Pheromones" (M. C. Birch, ed.), pp. 135–160. Am. Elsevier, New York.

90. Borden, J. H., and Bennett, R. B. (1969). A continuously recording flight mill for investigating the effect of volatile substances on the flight of tethered insects. *J. Econ. Entomol.* **62,** 782–785.

91. Borden, J. H., and Stokkink, E. (1973). Laboratory investigation of secondary attraction in *Gnathotrichus sulcatus* (Coleoptera:Scolytidae). *Can. J. Zool.* **51,** 469–473.

92. Bossert, W. H., and Wilson, E. O. (1963). The analysis of olfactory communication among animals. *J. Theor. Biol.* **5,** 443–469.

93. Bourlière, F. (1954). "The Natural History of Mammals." Alfred A. Knopf, New York.

94. Bowers, W. S., Nault, L. R. Webb, R. E., and Dutky, S. R. (1972). Aphid alarm pheromone: Isolation, identification, synthesis. *Science* **177,** 1121–1122.

95. Brader-Breukel, L. M. (1969). Modalités de l'attraction sexuelle chez *Diparopsis watersi* (Roths.). *Coton Fibres Trop. (Fr. Ed.)* **23, 24, 25.**

96. Bradley, J. R., Jr., Clower, D. F., and Graves, J. B. (1968). Field studies of sex attraction in the boll weevil. *J. Econ. Entomol.* **61,** 1457–1458.

97. Brady, U. E. (1973). Isolation, identification and stimulatory activity of a second component of the sex pheromone system (complex) of the female moth, *Cadra cautella* (Walker). *Life Sci.* **13,** 227–235.

98. Brady, U. E., and Daley, R. C. (1972). Identification of a sex pheromone from the female raisin moth, *Cadra figulilella. Ann. Entomol. Soc. Am.* **65,** 1356–1358.

99. Brady, U. E., and Ganyard, M. C., Jr. (1972). Identification of a sex pheromone of the female beet armyworm, *Spodoptera exigua. Ann. Entomol. Soc. Am.* **65,** 898–899.

100. Brady, U. E., and Nordlund, D. A. (1971). *Cis*-9, *trans*-12 tetradecadien-l-yl acetate in the female tobacco moth *Ephestia elutella* (Hübner) and evidence for an additional component of the sex pheromone. *Life Sci.* **10**, 797–801.

101. Brady, U. E., and Smithwick, E. B. (1968). Production and release of sex attractant by the female Indian-meal moth, *Plodia interpunctella. Ann. Entomol. Soc. Am.* **61**, 1260–1265.

102. Brady, U. E., Nordlund, D. A., and Daley, R. C. (1971). The sex stimulant of the Mediterranean flour moth, *Anagasta kuehniella. J. Ga. Entomol. Soc.* **6**, 215–217.

103. Brady, U. E., Tumlinson, J. H., III, Brownlee, R. G., and Silverstein, R. M. (1971). Sex stimulant and attractant in the Indian meal moth and in the almond moth. *Science* **171**, 802–804.

104. Bragg, D. E. (1974). Ecological and behavioral studies of *Phaeogenes cynarae:* Ecology; host specificity; searching and oviposition, and avoidance of superparasitism. *Ann. Entomol. Soc. Am.* **67**, 931–936.

105. Bronson, F. H. (1974). Pheromonal influences on reproductive activities in rodents. *In* "Pheromones" (M. C. Birch, ed.), pp. 344–365. Am. Elsevier, New York.

106. Bronson, F. H., and Marsden, H. M. (1973). The preputial gland as an indicator of social dominance in male mice. *Behav. Biol.* **9**, 625–628.

107. Brooksbank, B. W. L., and Hazelwood, G. A. D. (1950). The nature of pregnadiol-like glucoside. *J. Biochem. (Tokyo)* **47**, 36–43.

108. Brossut, R., Dubois, P., and Rigaud, J. (1974). Le grégarisme chez *Blaberus craniifer*: Isolement et identification de la phéromone. *J. Insect Physiol.* **20**, 529–543.

109. Brower, L. P., and Jones, M. A. (1965). Precourtship interaction of wing and abdominal sex glands in male *Danaus* butterflies. *Proc. R. Entomol. Soc. London (A)* **40**, 147–151.

110. Brower, L. P., Brower, J. V. Z., and Cranston, F. P. (1965). Courtship behavior of the queen butterfly, *Danaus gilippus berenice* (Cramer). *Zoologica (N.Y.)* **50**, 1–40.

111. Brown, L. N. (1972). Mating behavior and life habits of the sweet-bay silk moth (*Callosamia carolina*). *Science* **176**, 73–75.

112. Brown, R. E. (1974). Sexual arousal, the Coolidge effect and dominance in the rat (*Rattus norvegicus*). *Anim. Behav.* **22**, 634–637.

113. Brownlee, R. G., Silverstein, R. M., Müller-Schwarze, D., and Singer, A. G. (1969). Isolation, identification and function of the chief component of the male tarsal scent in black-tailed deer. *Nature (London)* **221**, 284–285.

114. Buechner, H. K., and Schloeth, R. (1965). Ceremonial mating behaviour in Uganda Kob (*Adenota kob thomasi* Neumann). *Z. Tierpsychol.* **22**, 209–225.

115. Burghardt, G. M. (1970). Chemical perception in reptiles. *Adv. Chemoreception* **1**, 241–308.

116. Burk, T., and Bell, W. J. (1973). Cockroach aggregation pheromone: Inhibition of locomotion (Orthoptera:Blattidae). *J. Kans. Entomol. Soc.* **46**, 36–41.

117. Butenandt, A., Beckmann, R., Stamm, D., and Hecker, E. (1959). Über den Sexuallockstoff des Seidenspinners *Bombyx mori* Reindarstellung und Konstitution. *Z. Naturforsch., Teil B* **14**, 283–384.

118. Butler, C. G. (1967). A sex attractant acting as an aphrodisiac in the honeybee (*Apis mellifera* L.). *Proc. R. Entomol. Soc. London (A)* **42**, 71–76.

119. Butler, C. G. (1970). Chemical communication in insects: Behavioral and ecologic aspects. *Adv. Chemoreception* **1**, 35–78.

120. Butler, C. G., and Fairey, E. M. (1964). Pheromones of the honeybee: Biological studies of the mandibular gland secretion of the queen. *J. Apic. Res.* **3**, 65–76.

121. Butler, C. G., and Simpson, J. (1967). Pheromones of the queen honeybee (*Apis*

mellifera L.) which enable her workers to follow her when swarming. *Proc. R. Entomol. Soc. London (A)* **42,** 149–154.

122. Butler, C. G., Callow, R. K., and Chapman, J. R. (1964). 9-Hydroxydec-*trans*-2-enoic acid, a pheromone stablizing honeybee swarms. *Nature (London)* **201,** 733.

123. Butler, C. G., Fletcher, D. J. C., and Walter, D. (1969). Nest-entrance marking with pheromones by the honeybee, *Apis mellifera* L., and by a wasp, *Vespula vulgaris* L. *Anim. Behav.* **17,** 142–147.

124. Calam, D. H., and Youdeowei, A. (1968). Identification and functions of secretion from the posterior scent gland of fifth instar larvae of the bug *Dysdercus intermedius*. *J. Insect Physiol.* **14,** 1147–1158.

125. Callow, R. K., and Johnston, N. C. (1960). The chemical constitution and synthesis of queen substance of honeybees (*Apis mellifera*). *Bee World* **41,** 152–153.

126. Cammaerts-Tricot, M. C. (1973). Phéromones agrégeant les ouvrières de *Myrmica rubra*. *J. Insect Physiol.* **19,** 1299–1315.

127. Cardé, R. T., and Roelofs, W. L. (1973). Temperature modification of male sex pheromone response and factors affecting female calling in *Holomelina immaculata* (Lepidoptera:Arctiidae). *Can. Entomol.* **105,** 1505–1512.

128. Carlson, D. A., Mayer, M. S., Silhacek, D. L., James, J. D., Beroza, M., and Bierl, B. A. (1971). Sex attractant pheromone of the house fly: Isolation, identification and synthesis. *Science* **174,** 76–78.

129. Caroom, D., and Bronson, F. H. (1971). Responsiveness of female mice to preputial attractant: Effects of sexual experience and ovarian hormones. *Physiol. Behav.* **7,** 659–662.

130. Carr, W. J., and Caul, W. F. (1962). The effect of castration in the rat upon the discrimination of sex odors. *Anim. Behav.* **10,** 20–27.

131. Carr, W. J., Loeb, L. S., and Dissinger, M. L. (1965). Responses of rats to sex odors. *J. Comp. Physiol. Psychol.* **59,** 370–377.

132. Casida, J. E., Coppel, H. C., and Watanabe, T. (1963). Purification and potency of the sex attractant from the introduced pine sawfly, *Diprion similis*. *J. Econ. Entomol.* **56,** 18–24.

133. Chararas, C. (1966). Recherches sur l'attractivité chez les Scolytidae. Etude sur l'attractivité sexuelle chez *Carphoborus minimum* Fabr. Coléoptère Scolytidae typiquement polygame. *C. R. Hebd. Seances Acad. Sci., Ser. D* **262,** 2492–2495.

134. Chararas, C. (1968). Recherches sur le comportement sexuel de *Pityokteines spinidens* Reit. (Coléoptère Scolytidae polygame) et étude des facteurs qui agissent sur le pouvoir attractif du mâle à l'égard de la femelle. *C. R. Hebd. Seances Acad. Sci., Ser. D* **266,** 1852–1855.

135. Chararas, C. (1969). Recherches sur l'attractivité sexuelle de *Phoceosinus bicolor* Brul., Coléoptère Scolytidae parasité specifique des Cupreisinae. *C. R. Hebd. Seances Acad. Sci., Ser. D* **268,** 1080–1083.

136. Chararas, C. (1971). L'intervention des facteurs nutritionnels dans la maturation et l'élaboration des phérhormones chez divers Scolytidae (Insects Coléoptères). *C. R. Hebd. Seances Acad., Sci., Ser. D* **272,** 2928–2931.

137. Chararas, C., and M'Sadda, K. (1970). Attraction chimique et attraction sexuelle chez *Orthotomicus erosus* Woll. (Coléoptèra:Scolytidae). *C. R. Hebd. Seances Acad. Sci., Ser. D* **271,** 1904–1907.

138. Cheng, R., and Samoiloff, M. R. (1971). Sexual attraction in the free-living nematode *Panagrellus silusiae* (Cephalobidae). *Can. J. Zool.* **49,** 1443–1448.

139. Chin, D. A., and Taylor, D. P. (1969). Sexual attraction and mating patterns in *Cylindrocorpus longistoma* and *C. curzii* (Nematoda:Cylindrocorporidae). *J. Nematol.* **1,** 313–317.

140. Clearwater, J. R. (1972). Chemistry and function of a pheromone produced by the male of the southern armyworm *Pseudaletia separata*. *J. Insect Physiol.* **18**, 781–789.

141. Coe, W. R. (1953). Influence of association, isolation, and nutrition on sexuality of snails in the genus *Crepidula*. *J. Exp. Zool.* **122**, 5–19.

142. Coffelt, J. A., and Burkholder, W. E. (1972). Reproductive biology of the cigarette beetle, *Lasioderma serricorne*. 1. Quantitative laboratory bioassay of the female sex pheromone from females of different ages. *Ann. Entomol. Soc. Am.* **65**, 447–450.

143. Cole, L. R. (1970). Observations on the finding of mates by male *Phaeogenes invisor* and *Apanteles medicaginis* (Hymenoptera:Ichneumonoidea). *Anim. Behav.* **18**, 184–189.

144. Collins, C. W., and Potts, S. F. (1932). Attractants for the flying gipsy moths as an aid in locating new infestations. *U.S., Dept. Agr., Tech. Bull.* **336**, 1–43.

145. Comeau, A., and Roelofs, W. L. (1973). Sex attraction specificity in the Tortricidae. *Entomol. Exp. Appl.* **16**, 191–200.

146. Comfort, A., (1971). Likelihood of human pheromones. *Nature (London)* **230**, 432–433.

147. Comfort, A. (1974). The likelihood of human pheromones. *In* "Pheromones" (M. C. Birch, ed.), pp. 386–396. Elsevier, New York.

148. Cone, W. W., and Pruszynski, S. (1972). Pheromone studies of the twospotted spider mite. 3. Response of males to different plant tissues, age, searching area, sex ratios, and solvents in bioassay trials. *J. Econ. Entomol.* **65**, 74–77.

149. Cone, W. W., McDonough, L. M., Maitlen, J. C., and Burdajewicz, S. (1971). Pheromone studies of the twospotted spider mite. *J. Econ. Entomol.* **64**, 355–358.

150. Cone, W. W., Predki, S., and Klostermeyer, E. C. (1971). Pheromone studies of the twospotted spider mite. 2. Behavioral responses of males to quiescent deutonymphs. *J. Econ. Entomol.* **64**, 379–382.

151. Cook, A., Bamford, O. S., Freeman, J. D. B., and Teideman, D. J. (1969). A study of the homing habit of the limpet. *Anim. Behav.* **17**, 330–339.

152. Corbet, S. A. (1971). Mandibular gland secretion of larvae of the flour moth, *Anagasta kuehniella*, contains an epideictic pheromone and elicits oviposition movements in a hymenopteran parasite. *Nature (London)* **232**, 481.

153. Cowan, B. D., and Rogoff, W. M. 1968. Variation and heritability of responsiveness of individual male house flies, *Musca domestica*, to the female sex pheromone. *Ann. Entomol. Soc. Am.* **61**, 1215–1218.

154. Cowley, J. J., and Wise, D. R. (1970). Pheromones, growth and behaviour. *Ciba Found. Study Group* **35**, 144–170.

155. Crewe, R. M., and Blum, M. S. (1972). Alarm pheromones of the Attini: Their phylogenetic significance. *J. Insect Physiol.* **18**, 31–42.

156. Crisp, D. J., and Meadows, P. S. (1962). The chemical basis of gregariousness in cirrepedes. *Proc. R. Soc. London, Ser. B* **156**, 500–520.

157. Crisp, D. J., and Meadows, P. S. (1963). Adsorbed layers: The stimulus to settlement in barnacles. *Proc. R. Soc. London, Ser. B* **158**, 364–387.

158. Crisp, M. (1969). Studies on the behaviour of *Nassarius obsoletus* (Say) (Mollusca, Gastropoda). *Biol. Bull.* **136**, 355–373.

159. Dadd, R. H., and Kleinjan, J. E. (1974). Autophagostimulant from *Culex pipiens* larvae: Distinction from other mosquito larval factors. *Environ. Entomol.* **3**, 21–28.

160. Dahm, K. H., Richter, I., Meyer, D., and Röller, H. (1971). The sex attractant of the Indian-meal moth, *Plodia interpunctella* (Hübner). *Life Sci., Part 2* **10**, 531–539.

161. Dahm, K. H., Meyer, D., Finn, W. E., Reinhold, V., and Röller, H. (1971). The olfactory and auditory mediated sex attraction in *Achroia grisella* (Fabr.). *Naturwissenschaften* **58**, 625–266.

162. Daterman, G. E. (1972). Laboratory bioassay for sex pheromone of the European pine shoot moth, *Rhyacionia buoliana. Ann. Entomol. Soc. Am.* **65**, 119–123.

163. Davies, V. J., and Bellamy, D. (1972). The olfactory response of mice to urine and effects of gonadectomy. *J. Endocrinol.* **55**, 11–20.

164. Davies, V. J., and Bellamy, D. (1974). Effects of female urine on social investigation in male mice. *Anim. Behav.* **22**, 239–241.

165. Dethier, V. G., Barton Browne, L., and Smith, C. N. (1960). The designation of chemicals in terms of the responses they elicit from insects *J. Econ. Entomol.* **53**, 134–136.

166. Devor, M., and Schneider, G. E. (1974). Attraction to home-cage odor in hamster pups: Specificity and changes with age. *Behav. Biol.* **10**, 211–221.

167. Diesendorf, M., Stange, G., and Snyder, A. W. (1974). A theoretical investigation of radiation mechanisms of insect chemoreception. *Proc. R. Soc. London, Ser. B* **185**, 33–49.

168. Dieterlen, F. (1959). Das Verhalten des syrischen Goldhamsters (*Mesocricetus auratus* Waterhouse). *Z. Tierpsychol.* **16**, 47–103.

169. Dinter, I. (1974). Pheromonal behavior in the marine snail *Littorina littorea* Linnaeus. *Veliger* **17**, 37–39.

170. Dixon, A. K., and Mackintosh, J. H. (1971). Effects of female urine upon the social behaviour of adult male mice. *Anim. Behav.* **19**, 138–140.

171. Doane, C. C. (1968). Aspects of mating behavior of the gypsy moth. *Ann. Entomol. Soc. Am.* **61**, 768–773.

172. Doane, C. C., and Cardé, R. T. (1973). Competition of gypsy moth males at a sex-pheromone source and a mechanism for terminating searching behavior. *Environ. Entomol.* **2**, 603–605.

173. Dobroruka, L. J. (1960). Einige Beobachtungen an Ameisenigeln, *Echidna aculeata* Shaw (1972). *Z. Tierpsychol.* **17**, 178–181.

174. Dondale, C. D., and Hegdekar, B. M. (1973). The contact sex pheromone of *Pardosa lapidicina* (Araneida:Lycosidae). *Can. J. Zool.* **51**, 400–401.

175. Doving, K. B., Nordeng, H., and Oakley, B. (1974). Single unit discrimination of fish odours released by char (*Salmo alpinus* L.) populations. *Comp. Biochem. Physiol. A* **47**, 1051–1063.

176. Downes, J. A. (1966). Observations on the mating behaviour of the crab hole mosquito *Deinocerites cancer* (Diptera:Culicidae). *Can. Entomol.* **98**, 1169–1177.

177. Doyle, G. A., Pelletier, A., and Bekker, T. (1967). Courtship, mating and parturition in the lesser bushbaby. *Folia Primatol.* **7**, 169–197.

178. Drickamer, L. C., Vandenbergh, J. G., and Colby, D. R. (1973). Predictors of dominance in the male golden hamster (*Mesocricetus auratus*). *Anim. Behav.* **21**, 557–563.

179. Dryden, G. L., and Conaway, C. H. (1967). The origin and hormonal control of scent production in *Suncus murinus. J. Mammal.* **48**, 420–428.

180. Dumais, J., Perron, J. -M., and Dondale, C. D. (1973). Elements du comportement sexuel chez *Pardosa xerampelina* (Keyserling) (Araneida:Lycosidae). *Can. J. Zool.* **51**, 265–271.

181. Dundee, H. A., and Miller, M. C., III. (1968). Aggregative behavior and habitat conditioning by the ringneck snake, *Diadophis punctatus arnyi. Tulane Stud. Zool. Bot.* **15**, 41–58.

182. Dyer, E. D. A., and Taylor, D. W. (1968). Attractiveness of logs containing female

spruce beetles, *Dendroctonus obesus* (Coleoptera:Scolytidae). *Can. Entomol.* **100**, 769–776.

183. Ehrman, L. (1972). A factor influencing the rare male mating advantage in *Drosophila. Behav. Gen.* **2**, 69–78.

184. Eibl-Eibesfeldt, I. (1964). Das Duftmarkieren des Igeltanrek (*Echinops telfairi* Martin). *Z. Tierpsychol.* **22**, 810–812.

185. Eisner, T. E., and Kafatos, F. C. (1962). Defense mechanisms of Arthropods. X. A pheromone promoting aggregation in an aposematic distasteful insect. *Psyche* **69**, 53–61.

186. Epple, G. (1974). Olfactory communication in South American primates. *Ann. N.Y. Acad. Sci.* **237**, 261–278.

187. Epple, G. (1974). Primate pheromones. *In* "Pheromones" (M. C. Birch, ed.), pp. 366–385. Am. Elsevier, New York.

188. Estes, R. D. (1972). The role of the vomeronasal organ in mammalian reproduction. *Mammalia* **36**, 315–341.

189. Evans, C. S., and Goy, R. W. (1968). Social behavior and reproductive cycles in captive ring-tailed lemurs (*Lemur catta*). *J. Zool.* **156**, 181–197.

190. Evans, K. (1970). Longevity of males and fertilisation of females of *Heterodera rostochiensis. Nematologica* **16**, 369–374.

191. Fales, H. M., Blum, M. S., Crewe, R. M., and Brand, J. M. (1972). Alarm pheromones in the genus *Manica* derived from the mandibular gland. *J. Insect Physiol.* **18**, 1077–1088.

192. Farkas, S. R., and Shorey, H. H. (1972). Chemical trail-following by flying insects: A mechanism for orienting to a distant odor source. *Science* **178**, 67–68.

193. Farkas, S. R., and Shorey, H. H. (1973). Odor-following and anemotaxis. *Science* **180**, 1302.

194. Farkas, S. R., and Shorey, H. H. (1974). Mechanisms of orientation to a distant pheromone source. *In* "Pheromones" (M. C. Birch, ed.), pp. 81–95. Am. Elsevier, New York.

195. Farkas, S. R., Shorey, H. H., and Gaston, L. K. (1974). Sex pheromones of Lepidoptera. Influence of pheromone concentration and visual cues on aerial odor-trail following by males of *Pectinophora gossypiella. Ann. Entomol. Soc. Am.* **67**, 633–638.

196. Fatzinger, C. W. (1972). Bioassay, morphology, and histology of the female sex pheromone gland of *Dioryctria abietella* (Lepidoptera:Pyralidae (Phycitinae)). *Ann. Entomol. Soc. Am.* **65**, 1208–1214.

197. Fatzinger, C. W. (1973). Circadian rhythmicity of sex pheromone release by *Dioryctria abietella* (Lepidoptera:Pyralidae (Phycitinae)) and the effect of a diel light cycle on its precopulatory behavior. *Ann. Entomol. Soc. Am.* **66**, 1147–1153.

198. Finley, H. E. (1952). Sexual differentiation in peritrichous ciliates. *J. Morphol.* **91**, 569–605.

199. Fletcher, B. S. (1969). The structure and function of the sex pheromone glands of the male Queensland fruit fly, *Dacus tryoni. J. Insect Physiol.* **15**, 1309–1322.

200. Fletcher, B.S., and Giannakakis, A. (1973). Factors limiting the response of females of the Queensland fruit fly, *Dacus tryoni*, to the sex pheromone of the male. *J. Insect Physiol.* **19**, 1147–1155.

201. Fletcher, D. J. C. (1971). The glandular source and social functions of trail pheromones in two species of ants (Leptogenys). *J. Entomol., Ser. A* **46**, 27–37.

202. Fraenkel, G. S., and Gunn, D. L. (1961). "The Orientation of Animals." Dover, New York.

203. Frank, A. (1941). Eigenartige Flugbahnen bei Hummelmännchen. *Z. Vergl. Physiol.* **28**, 467–484.

204. Free, J. B. (1961). The stimuli releasing the stinging response of honey bees. *Anim. Behav.* **9**, 193–196.

205. Free, J. B., and Williams, I. H. (1972). The role of the Nassonov gland pheromone in crop communication by honeybees (*Apis mellifera* L.). *Behaviour* **41**, 314–318.

206. Friedlander, C. P. (1965). Aggregation in *Oniscus ascellus* Linn. *Anim. Behav.* **13**, 342–346.

207. Furniss, M. M., and Schmitz, R. F. (1971). Comparative attraction of Douglas-fir beetles to frontalin and tree volatiles. *U.S., For. Serv., Res. Pap. Int.* **96**, 1–16.

208. Furniss, M. M., Kline, L. N., Schmitz, R. F., and Rudinsky, J. A. (1972). Tests of three pheromones to induce or disrupt aggregation of Douglas-fir beetles (Coleoptera:Scolytidae) on live trees. *Ann. Entomol. Soc. Am.* **65**, 1227–1232.

209. Gabba, A., and Pavan, M. (1970). Researches on trail and alarm substances in ants. *Adv. Chemoreception* **1**, 161–203.

210. Ganyard, M. C., Jr., and Brady, U. E. (1971). Inhibition of attraction and cross-attraction by interspecific sex pheromone communication in Lepidoptera. *Nature (London)* **234**, 415–416.

211. Ganyard, M. C., Jr., and Brady, U. E. (1972). Interspecific attraction in Lepidoptera in the field. *Ann. Entomol. Soc. Am.* **65**, 1279–1282.

212. Gara, R. I., and Coster, J. E. (1968). Studies on the attack behavior of the southern pine beetle. III. Sequence of tree infestation within stands. *Contrib. Boyce Thompson Inst.* **24**, 77–86.

213. Gara, R. I., Vité, J. P., and Cramer, H. H. (1965). Manipulation of *Dendroctonus frontalis* by use of a population aggregating pheromone. *Contrib. Boyce Thompson Inst.* **23**, 55–66.

214. Gary, N. E. (1962). Chemical mating attractants in the queen honey bee. *Science* **136**, 773–774.

215. Gary, N. E. (1970). Pheromones of the honey bee. *In* "Control of Insect Behavior by Natural Products" (D. L. Wood, R. M. Silverstein, and M. Nakajima, eds.), pp. 29–53. Academic Press, New York.

216. Gary, N. E. (1974). Pheromones that affect the behavior and physiology of honey bees. *In* "Pheromones" (M. C. Birch, ed.), pp. 200–201. Am. Elsevier, New York.

217. Gaston, L. K., Payne, T. L., Takahashi, S., and Shorey, H. H. (1972). Correlation of chemical structure and sex pheromone activity in *Trichoplusia ni* (Noctuidae). *In* "Olfaction and Taste IV," pp. 167–173. Wiss. Verlagsges., Stuttgart.

218. Gehlbach, F. R., Watkins, J. F., and Knoll, J. C. (1971). Pheromone trail following studies of typhlophid, leptotyphlophid, and colubrid snakes. *Behaviour* **40**, 282–294.

219. Geist, V. (1964). On the rutting behavior of the mountain goat. *J. Mammal.* **45**, 551–568.

220. George, J. A. (1965). Sex pheromone of the oriental fruit moth *Grapholitha molesta* – Busck (Lepidoptera, Tortricidae). *Can. Entomol.* **97**, 1002–1007.

221. Ghent, R. L. (1961). Adaptive refinements in the chemical defensive mechanisms of certain Formicinae. Ph.D. Thesis, Cornell University, Ithaca, New York.

222. Gilbert, J. J. (1963). Contact chemoreception, mating behavior and sexual isolation in the rotifer genus *Brachionus*. *J. Exp. Biol.* **40**, 625–641.

223. Gladney, W. J., Grabbé, R. R., Ernst, S. E., and Oehler, D. D. (1974). The Gulf Coast tick; evidence of a pheromone produced by males. *J. Med. Entomol.* **11**, 303–306.

224. Gladney, W. J., Ernst, S. E., and Grabbé, R. R. (1974). The aggregation response of the Gulf Coast tick on cattle. *Ann. Entomol. Soc. Am.* **67**, 750–752.

225. Goddard, J. (1967). Home range, behavior, and recruitment rates of two black rhinoceros populations. *East Afr. Wildl. J.* **5,** 133–150.

226. Godfrey, J. (1958). The origin of sexual isolation between bank voles. *Proc. R. Phys. Soc. Edinburgh* **27,** 47–55.

227. Golley, F. B. (1957). Gestation period, breeding and fawning behavior of Columbian black-tailed deer. *J. Mammal.* **38,** 116–120.

228. Goodrich, B. S., and Mykytowycz, R. (1972). Individual and sex differences in the chemical composition of pheromone-like substances from the skin glands of the rabbit, *Oryctolagus cuniculus. J. Mammal.* **53,** 540–548.

229. Goonewardene, H. F., Zepp, D. B., and Grosvener, A. E. (1970). Virgin female Japanese beetles as lures in field traps. *J. Econ. Entomol.* **63,** 1001–1003.

230. Gothilf, S., and Shorey, H. H. (1975). Sex pheromones of Lepidoptera: Examination of the role of male scent brushes in courtship behavior of *Trichoplusia ni. Environ. Entomol.* **5,** 115–119.

231. Götz, B. (1951). Die Sexualduftstoffe an Lepidopteren. *Experientia* **7,** 406–418.

232. Gower, D. P. (1972). 16-Unsaturated C_{19} steroids; a review of their chemistry, biochemistry and possible physiological role. *J. Steroid Biochem.* **3,** 45–103.

233. Graf, W. (1956). Territorialism in deer. *J. Mammal.* **37,** 165–170.

234. Gray, B., Billings, R. F., Gara, R. I., and Johnsey, R. L. (1972). On the emergence and initial flight behaviour of the mountain pine beetle, *Dendroctonus ponderosae,* in eastern Washington. *Z. Angew. Entomol.* **71,** 250–259.

235. Green, C. D. (1966). Orientation of male *Heterodera rostochiensis* Woll. and *H. schachtii* Schm. to their females. *Ann. Appl. Biol.* **58,** 327–339.

236. Green, C. D. (1967). The attraction of male cyst-nematodes by their females. *Nematologica* **13,** 172–173.

237. Green, C. D., and Greet, D. N. (1972). The location of the secretions that attract male *Heterodera schachtii* and *H. rostochiensis* to their females. *Nematologica* **18,** 347–352.

238. Green, G. W. (1962). Flight and dispersal of the European pine shoot moth, *Rhyacionia buoliana* (Schiff). I. Factors affecting flight and the flight potential of females. *Can. Entomol.* **94,** 282–299.

239. Green, N., Jacobson, M., Henneberry, T. J., and Kishaba, A. N. (1967). Insect sex attractants. VI. 7-dodecen-1-ol acetates and congeners. *J. Med. Chem.* **10,** 533–535.

240. Greenburg, B. (1943). Social behavior of the western banded gecko, *Coleonyx variegatus* Baird. *Physiol. Zool.* **16,** 110–122.

241. Greet, D. N. (1964). Observations on sexual attraction and copulation in the nematode *Panagrolaimus rigidus* (Schneider). *Nature (London)* **204,** 96–97.

242. Greet, D. N., Green, C. D., and Poulton, M. E. (1968). Extraction, standardization and assessment of the volatility of the sex attractants of *Heterodera rostochiensis* Woll. and *H. schachtii* Schm. *Ann. Appl. Biol.* **61,** 511–519.

243. Grubb, T. C., Jr. (1973). Odor-following and anemotaxis. *Science* **180,** 1302.

244. Grubb, T. C. (1974). Olfactory navigation to the nesting burrow in Leach's petrel *(Oceanodroma leucorrhoa). Anim. Behav.* **22,** 192–202.

245. Guerra, A. A. (1968). New techniques to bioassay the sex attractant of pink bollworms with olfactometers. *J. Econ. Entomol.* **61,** 1252–1254.

246. Haacker, U. (1969). An attractive secretion in the mating behaviour of a millipede. *Z. Tierpsychol.* **26,** 988–990.

247. Haacker, U. (1971). Die Funktion eines dorsalen Druesenkomplexes im Balzverhalten von *Chorodeuma* (Diplopoda). *Forma Functio* **4,** 162–170.

248. Haas, A. (1946). Neue Beobachtungen zum Problem der Flugbahnen beim Hummelmännchen. *Z. Naturforsch.* **11,** 596–600.

249. Haas, A. (1949). Art-typische Flugbahnen von Hummelmännchen. *Z. Vergl. Physiol.* **31**, 281–302.

250. Haas, A. (1952). Mandibeldrüse als Duftorgan bei einigen Hymenopteren. *Z. Naturforsch., Teil B* **39**, 484.

251. Hafez, E., Sumption, L., and Jakway, J. (1962). The behaviour of swine. *In* "The Behaviour of Domestic Animals" (E. Hafez, ed.), pp. 334–369. Williams & Wilkins, Baltimore, Maryland.

252. Haldane, J. B. S. (1954). La signalisation animale. *Annee Biol.* **58**, 89–98.

253. Haldane, J. B. S. (1955). Animal communication and the origin of human language. *Sci. Prog. (Oxford)* **43**, 385–401.

254. Hangartner, W. (1967). Spezifität und Inaktierung des Spurpheromons von *Lasius fuliginosus* Latr. und Orientierung der Arbeiterinnen im Duftfeld. *Z. Vergl. Physiol.* **57**, 103–136.

255. Hangartner, W. (1969). Orientierung von *Lasius fuliginosus* Latr. an einer gabelung der geruchsspur. *Insectes Soc.* **16**, 55–60.

256. Hangartner, W., Reichson, J. M., and Wilson, E. O. (1970). Orientation to nest material by the ant, *Pogonomyrmex badius* (Latreille). *Anim. Behav.* **18**, 331–334.

257. Hannes, F. (1965). Über den sinnesphysiologischen Mechanismus beim Aufsuchen und Auffinden entfernter Duftquellen in strömender Luft durch fliegende Insekten im Gegensatz zu kriechenden. *Biol. Zentralbl.* **84**, 191–203.

258. Happ, G. M. (1960). Multiple sex pheromones of the mealworm beetle, *Tenebrio molitor* L. *Nature (London)* **222**, 180–181.

259. Happ, G. M., and Wheeler J. (1969). Bioassay, preliminary purification, and effect of age, crowding, and mating on the release of sex pheromone by female *Tenebrio molitor. Ann. Entomol. Soc. Am.* **62**, 846–851.

260. Hardee, D. D., McKibben, G. H., Rummel, D. R., Huddleston, P. M., and Coppedge, J. R. (1974). Response of boll weevils to component ratios and doses of the pheromone, grandlure. *Environ. Entomol.* **3**, 135–138.

261. Harrison, B. (1963). Trying to breed *Tarsius. Malay. Nat. J.* **17**, 218–231.

262. Hartman, B. H., and Suda, M. (1973). Pheromone production and mating behaviour by allatectomized males of the cockroach, *Nauphoeta cinerea. J. Insect Physiol.* **19**, 1417–1422.

263. Harvey, E. B., and Rosenberg, L. E. (1960). An apocrine gland complex of the pika. *J. Mammal.* **41**, 213–219.

264. Haug, M. (1970). Mise en évidence de deux odeurs aux effets opposés de facilitation et d'inhibitiono des conduites aggressives chez la souris male. *C. R. Hebd. Seances Acad., Sci., Ser. D* **271**, 1567–1570.

265. Haug, M. (1971). Rôle probable des vésicules séminales et des glandes coagulantes dans la production d'une pheromone inhibitrice du comportement agressif chez la souris. *C. R. Hebd. Seances Acad. Sci., Ser. D* **273**, 1509–1510.

266. Haug, M. (1972). Effet de l'urine d'une femelle étrangere sur le comportement agressif d'un groupe de souris femelles. *C. R. Hebd. Seances Acad. Sci., Ser. D* **275**, 995–998.

267. Hazlett, B. A. (1970). Tactile stimuli in the social behavior of *Pagurus bernhardus* (Decapoda:Paguridae). *Behaviour* **36**, 20–48.

268. Hediger, H. (1949). Säugtier Territorien und ihre Markierung. *Bijdr. Dierk.* **28**, 172–184.

269. Hegdekar, B. M., and Dondale, C. D. (1969). A contact sex pheromone and some response parameters in lycosid spiders. *Can. J. Zool.* **47**, 1–4.

270. Hemmings, C. C. (1966). Olfaction and vision in fish schooling. *J. Exp. Biol.* **45**, 449–464.

271. Henderson, H. E., Warren, F. L., Augustyn, O. P. H., Burger, B. V., Schneider, D. F., Boshoff, P. R., Spies, H. S. C., and Geertsema, H. (1972). Sex-pheromones: *cis*-dec-5-en-1-yl 3-methyl-butanoate as the pheromone from the pine emperor moth (*Nudaurelia cytherea cytherea* Fabr.). *J. Chem. Soc., Chem. Commun.* **11**, 486–487.

272. Hendry, L. B., Roman, L., and Mumma, R. O. (1973). Evidence of a sex pheromone in the oak leaf roller, *Archips semiferanus* (Lepidoptera:Tortricidae): Laboratory and field bioassays. *Environ. Entomol.* **2**, 1024–1028.

273. Henzell, R. F., and Lowe, M. D. (1970). Sex attractant of the grass grub beetle. *Science* **168**, 1005–1006.

274. Hermann, H. R., Moser, J. C., and Hunt, A. N. (1970). The hymenopterous poison apparatus. X. Morphological and behavioral changes in *Atta texana* (Hymenoptera:Formicidae). *Ann. Entomol. Soc. Am.* **63**, 1553–1558.

275. Hertel, G. D., Hain, E. P., and Anderson, R. F. (1969). Response of *Ips grandicollis* (Coleoptera:Scolytidae) to the attractant produced by attacking male beetles. *Can. Entomol.* **101**, 1084–1091.

276. Hesterman, E. R., and Mykytowycz, R. (1968). Some observations on the intensities of odors of anal gland secretions from the rabbit *Oryctolagus cuniculus* (L). *CSIRO Wildl. Res.* **13**, 71–81.

277. Hidaka, T. (1972). Biology of *Hyphantria cunea* Drury (Lepidoptera, Arctiidae) in Japan. XIV. Mating behavior. *Appl. Entomol. Zool.* **7**, 116–132.

278. Hill, A. S., and Roelofs, W. L. (1975). Sex pheromone components of the omnivorous leafroller moth, *Platynota stultana. J. Chem. Ecol.* **1**, 91–99.

279. Hill, A. S., Cardé, R., Comeau, A., Bode, W., and Roelofs, W. L. (1974). Sex pheromones of the tufted apple bud moth (*Platynota idaeusalis*). *Environ. Entomol.* **3**, 249–252.

280. Hodek, J. (1960). Hibernation—bionomics in Coccinellidae. *Acta Soc. Entomol. Cech.* **57**, 1–20.

281. Hoellbodler, B., Moeglich, M., and Maschwitz, U. (1974). Communication by tandem running in the ant *Camponotus sericeus. J. Comp. Physiol.* **90**, 105–127.

282. Hölldobler, B. (1971). Homing in the harvester ant *Pogonomyrmex badius. Science* **171**, 1149–1151.

283. Hölldobler, B., and Maschwitz, U. (1965). Der Hochzeitsschwarm der Rossameise *Camponotus herculeanus* L. (Hym. Formicidae). *Z. Vergl. Physiol.* **50**, 551–568.

284. Hölldobler, B., and Wilson, E. O. (1970). Recruitment trails in the harvester ant *Pogonomyrmex badius. Psyche* **77**, 385–399.

285. Hölldobler, B., and Wüst, M. (1973). Ein Sexualpheromon bei der Pharaoameise *Monomorium pharaonis* (L.). *Z. Tierpsychol.* **32**, 1–9.

286. Hsiao, S. (1965). Effect of female variation on sexual satiation in the male rat. *J. Comp. Physiol. Psychol.* **60**, 467–469.

287. Hudson, A., and McLintock, J. (1967). A chemical factor that stimulates oviposition by *Culex tarsalis* Coquillet (Diptera, Culicidae). *Anim. Behav.* **15**, 336–341.

288. Hughes, P. R. (1973). *Dendroctonus:* Production of pheromones and related compounds in response to host monoterpenes. *Z. Angew. Entomol.* **73**, 294–312.

289. Hughes, P. R., (1973). Effect of α-pinene exposure on *trans*-verbenol synthesis in *Dendroctonus ponderosae* Hopk. *Naturwissenschaften* **60**, 261–262.

290. Hughes, P. R. (1974). Myrcene: A precursor of pheromones in *Ips* beetles. *J. Insect Physiol.* **20**, 1271–1275.

291. Hummel, H., and Karlson, P. (1968). Hexansäure als Bestandteil des Spurpheromons der Termite *Zootermopsis nevadensis* (Hagen). *Hoppe-Seyler's Z. Physiol. Chem.* **349**, 725–727.

292. Hummel, H. E., Gaston, L. K., Shorey, H. H., Kaae, R. S., Byrne, K. J., and Silverstein, R. M. (1973). Clarification of the chemical status of the pink bollworm sex pheromone. *Science* **181**, 873–875.

293. Ignoffo, C. M., Berger, R. S., Graham, H. M., and Martin, D. F. (1963). Sex attractant of cabbage looper, *Trichoplusia ni* (Hubner). *Science* **141**, 902–903.

294. Ikan, R., Gottlieb, R., Bergmann, E. D., and Ishay, J. (1969). The pheromone of the queen of the Oriental hornet, *Vespa orientalis. J. Insect Physiol.* **15**, 1709–1712.

295. Ilse, D. R. (1955). Olfactory marking of territory in two young male loris, *Loris tardigradus lydekkerianus,* kept in captivity in Poona. *Br. J. Anim. Behav.* **3**, 118–120.

296. Ishay, I., Ikan, R., and Bergmann, E. D. (1965). The presence of pheromones in the oriental hornet, *Vespa orientalis* F. *J. Insect Physiol.* **11**, 1307–1309.

297. Ishii, S. (1970). An aggregation pheromone of the German cockroach: *Blatella germanica* (L.). II. Species specificity of the pheromone. *Appl. Entomol. Zool.* **5**, 33–41.

298. Ishii, S. (1970). Aggregation of the German cockroach, *Blatella germanica* (L). *In* "Control of Insect Behavior by Natural Products" (D. L. Wood, R. M. Silverstein, and M. Nakajima, eds.), pp. 93–109. Academic Press, New York.

299. Ishii, S., and Kuwahara, Y. (1967). An aggregation pheromone of the German cockroach *Blattella germanica* L. (Orthoptera:Blattellidae). *Appl. Entomol. Zool.* **2**, 203–217.

300. Ishii, S., and Kuwahara, Y. (1968). Aggregation of German cockroach (*Blattella germanica*) nymphs. *Experientia* **24**, 88–89.

301. Jacobson, M. (1972). "Insect Sex Pheromones." Academic Press, New York.

302. Jacobson, M., Lilly, C. E., and Harding, C. (1968). Sex attractant of sugar beet wireworm: Identification and biological activity. *Science* **159**, 208–210.

303. Jacobson, M., Toba, H. H., DeBolt, J., and Kishaba, A. N. (1968). Insect sex attractants. VIII. Structure-activity relationships in sex attractants for male cabbage loopers. *J. Econ. Entomol.* **61**, 84–85.

304. Jacobson, M., Redfern, R. E., Jones, W. A., and Aldridge, M. H. (1970). Sex pheromones of the southern armyworm moth: Isolation, identification, and synthesis. *Science* **170**, 542–543.

305. Jacobson, M., Ohinata, K., Chambers, D. L., Jones, W. A., and Fujimoto, M. S. (1973). Insect attractants. 13. Isolation, identification, and synthesis of sex pheromones of the male Mediterranean fruit fly. *J. Med. Chem.* **16**, 248–251.

306. Jaisson, P. (1972). Nouvelles expériences sur l'aggressivité chez les Forms: Existence probable d'une substance active inhibitrice de l'aggressivité et attractive secretée par la jeune Formicine. *C. R. Hebd. Seances Acad. Sci., Ser. D* **274**, 302–305.

307. Jaisson, P. (1972). Mise en évidence d'une phéromone d'attractivité produite par la jeune ouvrière *Formica* (Hymenoptera:Formicidae). *C. R. Hebd. Seances Acad. Sci., Ser. D* **274**, 429–432.

308. Jantz, O. K., and Rudinsky, J. A. (1965). Laboratory and field methods for assaying olfactory responses of the Douglas fir beetle, *Dendroctonus pseudotsugae* Hopkins. *Can. Entomol.* **97**, 935–941.

309. Johnson, R. P. (1973). Scent marking in mammals. *Anim. Behav.* **21**, 521–535.

310. Jolly, A. (1966). "Lemur Behavior: A Madagascar Field Study." Univ. of Chicago Press, Chicago, Illinois.

311. Jolly, A. (1967). Breeding synchrony in wild *Lemur catta. In* "Social Communication Among Primates" (S. A. Altman, ed.), pp. 3–14. Univ. of Chicago Press, Chicago, Illinois.

312. Jones, R. B., and Nowell, N. W. (1973). Aversive effects of the urine of a male

mouse upon the investigatory behaviour of its defeated opponent. *Anim. Behav.* **21,** 707–710.

313. Jones, R. B., and Nowell, N. W. (1973) . The coagulating glands as a source of aversive and aggression-inhibiting pheromone (s) in the male albino mouse. *Physiol. Behav.* **11,** 455–462.

314. Jones, R. B., and Nowell, N. W. (1974) . A comparison of the aversive and female attractant properties of urine from dominant and subordinate male mice. *Anim. Learn. Behav.* **2,** 141–144.

315. Jones, R. B., and Nowell, N. W. (1974) . Effects of androgen on the aversive properties of the male mouse urine. *J. Endocrinol.* **60,** 19–25.

316. Jones, R. B., and Nowell, N. W. (1974) . The urinary aversive pheromone of mice: Species, strain, and grouping effects. *Anim. Behav.* **22,** 187–191.

317. Jones, T. P. (1966) . Sex attraction and copulation in *Pelodera teres. Nematologica* **12,** 518–522.

318. Kaae, R. S., and Shorey, H. H. (1972) . Sex pheromones of noctuid moths. XXVII. Influence of wind velocity on sex pheromone releasing behavior of *Trichoplusia ni* females. *Ann. Entomol. Soc. Am.* **65,** 436–440.

319. Kaae, R. S., and Shorey, H. H. (1973) . Sex pheromones of Lepidoptera. 44. Influence of environmental conditions on the location of pheromone communication and mating in *Pectinophora gossypiella. Environ. Entomol.* **2,** 1081–1084.

320. Kaae, R. S., Shorey, H. H., and Gaston, L. K. (1973) . Pheromone concentration as a mechanism for reproductive isolation between two lepidopterous species. *Science* **179,** 487–488.

321. Kaae, R. S., Shorey, H. H., McFarland, S. U., and Gaston, L. K. (1973) . Sex pheromones of Lepidoptera. XXXVII. Role of sex pheromones and other factors in reproductive isolation among ten species of Noctuidae. *Ann. Entomol. Soc. Am.* **66,** 444–448.

322. Kaissling, K.-E., and Priesner, E. (1970) . The olfactory threshold of the silk moth. *Naturwissenschaften* **57,** 23–28.

323. Kalkowski, W. (1967) . Olfactory bases of social orientation in the white mouse. *Folia Biol. (Krakow)* **15,** 69–87.

324. Kalmus, H. (1965) . Possibilities and constraints of chemical telecommunication. *Proc. Int. Congr. Endocrinol., 2nd, 1964* Int. Congr. Ser. No. 83, pp. 188–192.

325. Kalpage, K. S. P., and Brust, R. A. (1973) . Oviposition attractant produced by immature *Aedes atropalpus. Environ. Entomol.* **2,** 729–730.

326. Kamiguchi, Y. (1972) . Mating behavior in the freshwater prawn, *Palaemon paucidens:* A study of the sex pheromone and its effect on males. *J. Fac. Sci., Hokkaido Univ. Ser. 6* **18,** 347–355.

327. Kamiguchi, Y. (1972) . A histological study of the "sternal gland" in the female freshwater prawn, *Palaemon paucidens,* a possible site of origin of the sex pheromone. *J. Fac. Sci., Hokkaido Univ., Ser. 6* **18,** 356–365.

328. Karandinos, M. G. (1974) . Environmental conditions and sex activity of *Synanthedon pictipes* in Wisconsin, monitored with virgin female pheromone traps. *Environ. Entomol.* **3,** 431–438.

329. Karlson, P., and Butenandt, A. (1959) . Pheromones (ectohormones) in insects. *Annu. Rev. Entomol.* **4,** 49–58.

330. Karlson, P., and Lüscher, M. (1959) . "Pheromones": A new term for a class of biologically active substances. *Nature (London)* **183,** 55–56.

331. Karlson, P., Lüscher, M., and Hummel, H. (1968) . Extraktion und Biologische Auswertung des Spurpheromons der Termite *Zootermopsis nevadensis. J. Insect Physiol.* **14,** 1763–1771.

332. Kaufman, T. (1966). Observations on some factors which influence aggregation by *Blaps sulcata* (Coleoptera:Tenebrionidae) in Israel. *Ann. Entomol. Soc. Am.* **59,** 660–664.

333. Keenleyside, M. H. S. (1955). Some aspects in the schooling behavior in fish. *Behaviour* **8,** 183–248.

334. Kellogg, F. E., Frizel, D. E., and Wright, R. H. (1962). The olfactory guidance of flying insects. IV. Drosophila. *Can. Entomol.* **94,** 884–888.

335. Kennedy, J. S. (1939). The visual responses of flying mosquitoes. *Proc. Zool. Soc. London, Ser. A* **109,** 221–242.

336. Kennedy, J. S., and Marsh, D. (1974). Pheromone-regulated anemotaxis in flying moths. *Science* **184,** 999–1001.

337. Kerr, W. E. (1960). Evolution of communication in bees and its role in speciation. *Evolution* **14,** 386–387.

338. Kettlewell, H. B. D. (1946). Female assembling scents with reference to an important paper on the subject. *Entomologist* **79,** 8–14.

339. Keverne, E. B., and Michael, R. P. (1971). Sex attractant properties of ether extracts of vaginal secretions from rhesus monkeys. *J. Endocrinol.* **51,** 313–322.

340. King, A. B. S. (1973). The actographic examination of flight activity of the cocoa mirid *Distantiella theobroma* (Hemiptera:Miridae). *Entomol. Exp. Appl.* **16,** 53–63.

341. King, A. B. S. (1973). Studies of sex attraction in the cocoa capsid, *Distantiella theobroma* (Heteroptera:Miridae). *Entomol. Exp. Appl.* **16,** 243–254.

342. Kingston, B. H. (1965). The chemistry and olfactory properties of musk, civet and castoreum. *Proc. Int. Congr. Endocrinol., 2nd, 1964* Int. Congr. Ser. No. 83, pp. 209–214.

343. Kinzer, G. W., Fentiman, A. F., Jr., Page, T. F., Jr., Foltz, R. L., Vité, J. P., and Pitman, G. B. (1969). Bark beetle attractants: Identification, synthesis and field bioassay of a new compound isolated from *Dendroctonus. Nature (London)* **221,** 447–478.

344. Kinzer, G. W., Fentiman, A. F., Jr., Foltz, R. L., and Rudinsky, J. A. (1971). Bark beetle attractants: 3-methyl-2-cyclohexen-1-one isolated from *Dendroctonus pseudotsugae. J. Econ. Entomol.* **64,** 970–971.

345. Kirk, V. M., and Dupraz, J. B. (1972). Discharge by a female ground beetle, *Pterostichus lucublandus* (Coleoptera:Carabidae), used as a defense against males. *Ann. Entomol. Soc. Am.* **65,** 513.

346. Kislow, C. J., and Edwards, L. J. (1972). Repellent odour in aphids. *Nature (London)* **235,** 108–109.

347. Kitamura, C., and Takahashi, S. (1973). The mating behavior and evidence for a sex stimulant of the Japanese cockroach, *Periplaneta japonica* Karny (Orthoptera: Blattidae). *Kontchu* **41,** 383–388.

348. Kittredge, J. S., and Takahashi, F. T. (1972). The evolution of sex pheromone communication in the Arthropoda. *J. Theor. Biol.* **35,** 467–471.

349. Kittredge, J. S., Terry, M., and Takahashi, F. T. (1971). Sex pheromone activity of the molting hormones, crustecdysone, on male crabs: (*Pachygrapsus crassipes, Cancer antennarius,* and *C. anthonyi*). *U.S., Fish Wild. Serv., Fish. Bull.* **69,** 337–343.

350. Klein, M. G., Ladd, T. L., Jr., and Lawrence, K. O. (1972). The influence of height of exposure of virgin female Japanese beetles on captures of males. *Environ. Entomol.* **1,** 600–601.

351. Klein, M. G., Ladd, T. L., Jr., and Lawrence, K. O. (1972). A field comparison of lures for Japanese beetles: Unmated females vs. phenethyl propionate + eugenol (7:3). *Environ. Entomol.* **1,** 397–399.

352. Kleinman, D. (1966). Scent marking in the Canidae. *Symp. Zool. Soc. London* **18**, 167–177.

353. Kliefoth, R. A., Vité, J. P., and Pitman, G. B. (1964). A laboratory technique for testing bark beetle attractants. *Contrib. Boyce Thompson Inst.* **22**, 283–290.

354. Kliewer, J. W., Miura, T., Husbands, R. C., and Hurst, C. H. (1966). Sex pheromones and mating behavior of *Culiseta inornata* (Diptera:Culicidae). *Ann. Entomol. Soc. Am.* **59**, 530–533.

355. Klun, J. A., and Brindley, T. A. (1970). *cis*-11-Tetradecenyl acetate, a sex stimulant of the European corn borer. *J. Econ. Entomol.* **63**, 779–780.

356. Klun, J. A., and Robinson, J. F. (1972). Olfactory discrimination in the European corn borer and several pheromonally analogous moths. *Ann. Entomol. Soc. Am.* **65**, 1337–1340.

357. Klun, J. A., Chapman, O. L., Mattes, K. C., Wojtkowski, P. W., Beroza, M., and Sonnet, P. E. (1973). Insect sex pheromones: Minor amount of opposite geometrical isomer critical to attraction. *Science* **181**, 661–663.

358. Kochansky, J., Cardé, R. T., Liebherr, J., and Roelofs, W. L. (1975). Sex pheromones of the European corn borer *(Ostrinia nubilalis)* in New York. *J. Chem. Ecol.* **1**, 225–231.

359. Koelega, H. S., and Köster, E. P. (1974). Some experiments on sex differences in odor perception. *Ann. N.Y. Acad. Sci.* **237**, 234–246.

360. Koenig, L. (1957). Beobachtungen über Reviermarkierung sowie Droh-, Kampf- und Abwehrverhalten des Murmeltieres *(Marmota marmota* L.). *Z. Tierpsychol.* **14**, 510–521.

361. Konijin, T. M., Barkley, D. S., Chang, Y. Y., and Bonner, J. T. (1968). Cyclic AMP: A naturally occurring acrasin in the cellular slime molds. *Am. Nat.* **102**, 225–233.

362. Kuenen, D. J., and Nooteboom, H. P. (1963). Olfactory orientation in some land-isopods (Oniscoidea, Crustacea). *Entomol. Exp. Appl.* **6**, 133–142.

363. Kühme, W. (1946). Chemisch ausgeloeste Brutpflege und Schwarmreaktionen bei *Hemichromis bimaculatus* (Pisces). *Z. Tierpsychol.* **20**, 688–704.

364. Kullenberg, B. (1956). Field experiments with chemical sexual attractants on Aculeate Hymenoptera males. *Zool. Bidr. Uppsala* **31**, 253–354.

365. Kuwahara, Y., Hara, H., Ishii, S., and Fukami, H. (1971). The sex pheromone of the Mediterranean flour moth. *Agr. Biol. Chem.* **35**, 447–448.

366. Kuwahara, Y., Kitamura, C., Takahashi, S., Hara, H., Ishii, S., and Fukami, H. (1971). Sex pheromone of the almond moth and the Indian meal moth: *cis*-9, *trans*-12-tetradecadienyl acetate. *Science* **171**, 801–802.

367. Lanier, G. N., and Burkholder, W. E. (1974). Pheromones in speciation of Coleoptera. *In* "Pheromones" (M. C. Birch, ed.), pp. 161–189. Am. Elsevier, New York.

368. LeComte, J. (1956). Über die Bildung von 'Strassen' durch Sammelbienen, deren stock um 180° gedrecht wurde. *Z. Bienenforsch.* **3**, 128–133.

369. Lederer, E. (1950). Odeurs et parfums des animaux. *Fortschr. Chem. Org. Naturst.* **6**, 87–153.

370. Lee, C. T., and Brake, S. C. (1971). Reactions of male fighters to male and female mice, untreated or deodorized. *Psychonomic Sci., Sect. Anim. Physiol. Psychol.* **24**, 209–211.

371. Lee, C. T., and Brake, S. C. (1972). Reaction of male mouse fighters to male castrates treated with testosterone proprionate or oil. *Psychonomic Sci., Sect. Anim. Physiol. Psychol.* **27**, 287–288.

372. Lee, C. T., and Griffo, W. (1972). Early androgenization and aggression pheromone in inbred mice. *Am. Zool.* (abstr.).

373. Lee, C. T., and Griffo, W. (1973). Early androgenization and aggression pheromone

in inbred mice. *Horm. Behav.* **4,** 181–189.

374. Le Magnen, J. (1948). Physiologie des sensations—un cas de sensibilité olfactaire se présentant comme un caractère sexuel secondaire féminin. *C. R. Hebd. Seances Acad. Sci.* **226,** 694–695.

375. Le Magnen, J. (1950). Physiologie des sensations—l'odeur des hormones sexuelles. *C. R. Hebd. Seances Acad. Sci.* **230,** 1367–1369.

376. Leon, M., and Moltz, H. (1971). Maternal pheromone: Discrimination by pre-weanling albino rats. *Physiol. Behav.* **7,** 265–267.

377. Leon, M., and Moltz, H. (1972). The development of the pheromonal bond in the albino rat. *Physiol. Behav.* **8,** 683–686.

378. Leon, M., and Moltz, H. (1973). Endocrine control of the maternal pheromone in the postpartum female rat. *Physiol. Behav.* **10,** 65–67.

379. Levinson, H. Z. (1972). Zur Evolution und Biosynthese der terpenoiden Pheromone und Hormone. *Naturwissenschaften* **59,** 477–484.

380. Levinson, H. Z., and Bar Ilan, A. R. (1967). Function and properties of an assembling scent in the Khapra beetle *Trogoderma granarium. Riv. Parassitol,* **28,** 27–42.

381. Levinson, H. Z., and Bar Ilan, A. R. (1970). Behaviour of khapra beetle *Trogoderma granarium* towards the assembling scent released by the female. *Experientia* **26,** 846–847.

382. Levinson, H. Z., and Bar Ilan, A. R. (1970). Lack of an intraspecific attractant in male *Trogoderma granarium. Riv. Parassitol.* **31,** 70–72.

383. Levinson, H. Z., and Bar Ilan, A. R. (1970). Olfactory and tactile behaviour of the khapra beetle, *Trogoderma granarium,* with special reference to its assembling scent. *J. Insect Physiol.* **16,** 561–572.

384. Levinson, H. Z., and Bar Ilan, A. R. (1971). Assembling and alerting scents produced by the bedbug *Cimex lectularius* L. *Experientia* **27,** 102–103.

385. Levinson, H. Z., Levinson, A. R., Müller, B., and Steinbrecht, R. A. (1974). Structure of sensilla, olfactory perception, and behaviour of the bedbug, *Cimex lectularius,* in response to its alarm pheromone. *J. Insect Physiol.* **20,** 1231–1248.

386. Leyrer, R. L., and Monroe, R. E. (1973). Isolation and identification of the scent of the moth, *Galleria mellonella,* and reevaluation of its sex pheromone. *J. Insect Physiol.* **19,** 2267–2271.

387. Libbey, L. M., Morgan, M. E., Putnam, T. B., and Rudinsky, J. A. (1974). Pheromones released during inter- and intra-sex response of the scolytid beetle *Dendroctonus brevicomis. J. Insect. Physiol.* **20,** 1667–1671.

388. Lindauer, M. (1967). Recent advances in bee communication and orientation. *Annu. Rev. Entomol.* **12,** 439–470.

389. Lindauer, M., and Kerr, W. E. (1958). Die gegenseitige Verständigung bei den stachellosen Bienen. *Z. Vergl. Physiol.* **41,** 405–434.

390. Lindauer, M., and Kerr, W. E. (1960). Communication between the workers of stingless bees. *Bee World* **41,** 29–41 and 65–71.

391. Lindauer, M., and Martin, H. (1963). Über die Orientierung der Biene im Duftfeld. *Naturwissenschaften* **15,** 509–514.

392. Lindsay, D. R. (1965). The importance of olfactory stimuli in the mating behaviour of the ram. *Anim. Behav.* **13,** 75–78.

393. Linsenmair, K. E., and Linsenmair, C. (1971). Paarbildung and Paarzusammenhalt bei der monogamen Wuestenassel *Hemilepistus reaumuri* (Crustacea, Isopoda, Oniscoidea). *Z. Tierpsychol.* **29,** 134–155.

394. Lloyd, J. E. (1972). Chemical communication in fireflies. *Environ. Entomol.* **1,** 265–266.

395. Lockie, J. D. (1966). Territory in small carnivores. *Symp. Zool. Soc. London* **18**, 143–165.

396. MacGinitie, G. E. (1939). The natural history of the blind goby, *Typhlogobius californiensis* (Steindachner). *Am. Midl. Nat.* **21**, 489–505.

397. Mackintosh, J. H. (1973). Factors affecting the recognition of territory boundaries by mice (*Mus musculus*). *Anim. Behav.* **21**, 464–470.

398. Mackintosh, J. H., and Grant, E. C. (1966). The effect of olfactory stimuli on the agonistic behaviour of laboratory mice. *Z. Tierpsychol.* **23**, 584–587.

399. Madrid, F., Vité, J. P., and Renwick, J. A. A. (1972). Evidence of aggregation pheromones in the ambrosia beetle *Platypus flavicornis* (F.). *Z. Angew. Entomol.* **72**, 73–79.

400. Marchant, H. V. (1970). Bursal response in sexually stimulated *Nematospiroides dubius* (Nematoda). *J. Parasitol.* **56**, 201–202.

401. Marler, P. (1965). Communication in monkeys and apes. *In* "Primate Behavior" (I. DeVore, ed), pp. 544–584. Holt, New York.

402. Marler, P., and Hamilton, W. J., III. (1967). "Mechanisms of Animal Behavior." Wiley, New York.

403. Martin, R. D. (1968). Reproduction and ontogeny in tree-shrews (*Tupaia belangeri*), with references to their general behaviour and taxonomic relationships. *Z. Tierpsychol.* **25**, 404–495, 505–532.

404. Maschwitz, U. W. (1964). Alarm substances and alarm behaviour in social Hymenoptera. *Nature (London)* **204**, 324–327.

405. Maschwitz, U. W. (1964). Gefahrenalarmstoffe und Gefahrenalarmierung bei sozialen Hymenopteren. *Z. Vergl. Physiol.* **47**, 596–655.

406. Maschwitz, U. W. (1966). Alarm substances and alarm behavior in social insects. *Vitam. Horm. (N.Y.)* **24**, 267–290.

407. Matthews, L. H. (1939). The bionomics of the spotted hyena, *Crocuta crocuta* Erxl. *Proc. Zool. Soc. London, Ser. A* **109**, 43–55.

408. Mautz, D., Boch, R., and Morse, R. A. (1972). Queen finding by swarming honey bees. *Ann. Entomol. Soc. Am.* **65**, 440–443.

409. McConnell, J. V. (1976). Specific factors influencing planarian behaviour. *In* "Symposium on Chemistry of Learning" (W. C. Corning and S. C. Ratner, eds.), pp. 217–233. Plenum, New York.

410. McLeese, D. W. (1973). Chemical communication among lobsters (*Homarus americanus*). *J. Fish Res. Board Can.* **30**, 775–778.

411. Meinwald, J., Meinwald, Y. C., and Mazzocchi, P. H. (1969). Sex pheromone of the queen butterfly: Chemistry. *Science* **164**, 1174–1175.

412. Melrose, D. R., Reed, H. C. B., and Patterson, R. L. S. (1971). Androgen steroids associated with boar odour as an aid to the detection of oestrus in pig artificial insemination. *Br. Vet. J.* **127**, 497–501

413. Michael, R. P. (1969). The role of pheromones in the communication of primate behaviour. *Recent Adv. Primatol.* **1**, 101–108.

414. Michael, R. P., and Keverne, E. B. (1968). Pheromones in the communication of sexual status in primates. *Nature (London)* **218**, 746–749.

415. Michael, R. P., and Keverne, E. B. (1970). Primate sex pheromones of vaginal origin. *Nature (London)* **225**, 84–85.

416. Michael, R. P., Keverne, E. B., and Bonsall, R. W. (1971). Pheromones: Isolation of male sex attractants from a female primate. *Science* **172**, 964–966.

417. Michael, R. P., Bonsall, R. W., and Warner, P. (1974). Human vaginal secretions: Volatile fatty acid content. *Science* **186**, 1217–1219.

418. Michael, R. R., and Rudinsky, J. A. (1972). Sound production in Scolytidae:

Specificity in male *Dendroctonus* beetles. *J. Insect Physiol.* **18**, 2189–2201.

419. Miller, C. A., and McDougall, G. A. (1973). Spruce budworm moth trapping using virgin females. *Can. J. Zool.* **51**, 853–858.

420. Minks, A. K., and Voerman, S. (1973). Sex pheromones of the summerfruit tortrix moth, *Adoxophyes orana:* Trapping performance in the field. *Entomol. Exp. Appl.* **16**, 541–549.

421. Minks, A. K., Roelofs, W. L., Ritter, F. J., and Persoons, C. J. (1973). Reproductive isolation of two tortricid moth species by different ratios of a two-component sex attractant. *Science* **180**, 1073–1074.

422. Mitchell, E. R. (1972). Female cabbage loopers inhibit attraction of male soybean loopers. *Environ. Entomol.* **1**, 444–446.

423. Mitchell, E. R., Hardee, D. D., Cross, W. H., Huddleston, P. M., and Mitchell, H. C. (1972). Influence of rainfall, sex ratio, and physiological condition of boll weevils on their response to pheromone traps. *Environ. Entomol.* **1**, 438–440.

424. Möglich, M., Maschwitz, U., and Hölldobler, B. (1974). Tandem calling: A new kind of signal in ant communication. *Science* **186**, 1046–1047.

425. Moltz, H., and Leon, M. (1973). Stimulus control of the maternal pheromone in the lactating rat. *Physiol. Behav.* **10**, 69–71.

426. Moltz, H., Leidahl, L., and Rowland, D. (1974). Prolongation of pheromonal emission in the maternal rat. *Physiol. Behav.* **12**, 409–412.

427. Moore, B. P. (1967). Chemical communication in insects. *Sci. J.* **3**, 44–49.

428. Moore, B. P. (1968). Studies on the chemical composition and function of the cephalic gland secretion in Australian termites. *J. Insect Physiol.* **14**, 33–39.

429. Moore, B. P. (1974). Pheromones in the termite societies. *In* "Pheromones" (M. C. Birch, ed.), pp. 250–266. Am. Elsevier, New York.

430. Morse, R. A. (1972). Honey bee alarm pheromone: Another function. *Ann. Entomol. Soc. Am.* **65**, 1430.

431. Morse, R. A., and Boch, R. (1971). Pheromone concert in swarming honey bees (Hymenoptera:Apidae). *Ann. Entomol. Soc. Am.* **64**, 1414–1417.

432. Morse, R. A., and Gary, N. E. (1963). Further studies of the responses of honey bee (*Apis mellifera* L.) colonies to queens with extirpated mandibular glands. *Ann. Entomol. Soc. Am.* **56**, 372–374.

433. Moser, J. C., and Blum, M. S. (1963). Trail marking substance of the Texas leaf-cutting ant: Source and potency. *Science* **140**, 1228.

434. Moser, J. C., Brownlee, R. C., and Silverstein, R. (1968). Alarm pheromones of the ant *Atta texana*. *J. Insect Physiol.* **14**, 529–535.

435. Mudd, A., and Corbet, S. A. (1973). Mandibular gland secretion of larvae of the stored products pest *Anagasta kuehniella, Ephestia cautella, Plodia interpunctella* and *Ephestia elutella. Entomol. Exp. Appl.* **16**, 291–293.

436. Mugford, R. A., and Nowell, N. W. (1970). Pheromones and their effect on aggression in mice. *Nature (London)* **226**, 967–968.

437. Mugford, R. A., and Nowell, N. W. (1971). The preputial glands as a source of aggression-promoting odors in mice. *Physiol. Behav.* **6**, 247–249.

438. Mugford, R. A., and Nowell, N. W. (1972). The dose-response to testosterone propionate of preputial glands, pheromones and aggression in mice. *Horm. Behav.* **3**, 39–46.

439. Müller, D. G., Jaenicke, L., Donike, M., and Akintobi, T. (1971). Sex attractant in a brown alga: Chemical structure. *Science* **171**, 815–817.

440. Müller-Schwarze, D. (1971). Pheromones in black-tailed deer (*Odocoileus hemionus columbianus*). *Anim. Behav.* **19**, 141–152.

441. Müller-Schwarze, D. (1972). Social significance of forehead rubbing in blacktailed

deer (*Odocoileus hemionus columbianus*). *Anim. Behav.* **20**, 788–797.

442. Müller-Schwarze, D. (1974). Olfactory recognition of species, groups, individuals and physiological states among mammals. *In* "Pheromones" (M. C. Birch, ed.), pp. 316–326. Am. Elsevier, New York.

443. Müller-Schwarze, D., Müller-Schwarze, C., Singer, A. G., and Silverstein, R. M. (1974). Mammalian pheromone: Identification of active component in the subauricular scent of the male pronghorn. *Science* **183**, 860–862.

444. Muller-Velten, H. (1966). Über den Angstgeruch bei der Hausmaus (*Mus musculus* L.). *Z. Vergl. Physiol.* **52**, 401–429.

445. Murphy, M. R. (1973). Effects of female hamster vaginal discharge on the behavior of male hamsters. *Behav. Biol.* **9**, 367–375.

446. Murphy, M. R., and Schneider, G. E. (1970). Olfactory bulb removal eliminates mating behavior in the male golden hamster. *Science.* **167**, 302–303.

447. Myers, J., and Brower, L. P. (1969). A behavioural analysis of the courtship pheromone receptors of the Queen butterfly, *Danaus gilippus berenice*. *J. Insect Physiol.* **15**, 2117–2130.

448. Mykytowycz, R. (1965). Further observations on the territorial function and histology of the submandibular cutaneous (chin) glands in the rabbit, *Oryctolagus cuniculus* (L.). *Anim. Behav.* **13**, 400–412.

449. Mykytowcyz, R. (1968). Territorial marking by rabbits. *Sci. Am.* **218**, 116–126.

450. Mykytowycz, R. (1970). The role of skin glands in mammalian communication. *Adv. Chemoreception* **1**, 327–360.

451. Mykytowycz, R., and Dudzinski, M. S. (1966). A study of the weight of odoriferous and other glands in relation to social status and degree of sexual activity in the wild rabbit, *Oryctolagus cuniculus* (L.). *CSIRO Wildl. Res.* **11**, 31–47.

452. Mykytowycz, R., and Gambale, S. (1969). The distribution of dunghills and the behavior of free-living rabbits, *Oryctolagus cuniculus* (L.), on them. *Forma Functio* **1**, 333–349.

453. Mykytowycz, R., and Hesterman, E. R. (1970). The behavior of captive wild rabbits, *Oryctolagus cuniculus* (L.), in response to strange dung-hills. *Forma Functio* **2**, 1–12.

454. Myrberg, A. A., Jr. (1966). Parental recognition of young in cichlid fishes. *Anim. Behav.* **14**, 565–571.

455. Nault, L. R., Edwards, L. J., and Styer, W. E. (1973). Aphid alarm pheromones: Secretion and reception. *Environ. Entomol.* **2**, 101–105.

456. Naylor, A. F. (1959). An experimental analysis of dispersal in the flour beetle, *Tribolium confusum*. *Ecology* **40**, 453–465.

457. Naylor, A. F. (1961). Dispersal in the red flour beetle *Tribolium castaneum* (Tenebrionidae). *Ecology* **42**, 231–237.

458. Naylor, A. F. (1965). Dispersal responses of female flour beetles, *Tribolium confusum*, to presence of larvae. *Ecology* **46**, 341–343.

459. Nesbitt, B. F., Beevor, P. S., Cole, R. A., Lester, R., and Poppi, R. G. (1973). Sex pheromones of two noctuid moths. *Nature (London)*, *New Biol.* **244**, 208–209.

460. Neumark, S., Jacobson, M., and Teich, I. (1974). Field evaluation of the four synthetic components of the sex pheromone of *Spodoptera littoralis* and their improvement with an antioxidant. *Environ. Lett.* **6**, 219–230.

461. Nielsen, D. G., and Balderston, C. P. (1973). Evidence for intergeneric sex attraction among aegeriids. *Ann. Entomol. Soc. Am.* **66**, 227–228.

462. Nijholt, W. W. (1970). The effect of mating and the presence of the male ambrosia beetle, *Trypodendron lineatum*, on "secondary" attraction. *Can. Entomol.* **102**, 894–897.

463. Nijholt, W. W. (1973). The effect of male *Trypodendron lineatum* (Coleoptera: Scolytidae) on the response of field populations to secondary attraction. *Can. Entomol.* **105**, 583–590.

464. Noble, G. K. (1937). The sense organs involved in the courtship of *Storeria, Thamnophis,* and other snakes. *Bull. Am. Mus. Nat. Hist.* **73**, 673–725.

465. Noble, G. K., and Clausen, H. J. (1936). The aggregation behavior of *Storeria dekayi* and other snakes with especial reference to the sense organs involved. *Ecol. Monogr.* **6**, 269–316.

466. Nolte, D. J., Eggers, S. H., and May, I. R. (1973). A locust pheromone:locustol. *J. Insect Physiol.* **19**, 1547–1554.

467. Nordeng, H. (1971). Is the local orientation of anadromous fishes determined by pheromones? *Nature (London)* **233**, 411–413.

468. Norris, M. J. (1963). Laboratory experiments on gregarious behaviour in ovipositing females of the desert locust, *(Schistocerca gregaria* (Forsk.)). *Entomol. Exp. Appl.* **6**, 279–303.

469. Norris, M. J. (1970). Aggregation response in ovipositing females of the desert locust, with special reference to the chemical factor. *J. Insect Physiol.* **16**, 1493–1515.

470. Ohbayashi, N., Yushima, T., Noguchi, H., and Tamaki, Y. (1973). Time of mating and sex pheromone production and release of *Spodoptera litura* (F.) (Lepidoptera:Noctuidae). *Konchu* **41**, 389–395.

471. Osgood, C. E. (1971). An oviposition pheromone associated with the egg rafts of *Culex tarsalis. J. Econ. Entomol.* **64**, 1038–1041.

472. Pain, J., and Barbier, M. (1960). Mise en évidence d'une substance attractive extraite du corps des ouvrières d'abeilles non orphelines *(Apis mellifera* L.). *C. R. Hebd. Seances Acad. Sci.* **250**, 1126–1127.

473. Payne, T. L. (1974). Pheromone perception. *In* "Pheromones" (M. C. Birch, ed.), pp. 35–61. Am. Elsevier, New York.

474. Payne, T. L., Shorey, H. H., and Gaston, L. K. (1970). Sex pheromones of noctuid moths: Factors influencing antennal responsiveness in males of *Trichoplusia ni. J. Insect Physiol.* **16**, 1043–1055.

475. Peacock, J. W., Lincoln, A. C., Simeone, J. B., and Silverstein, R. M. (1971). Attraction of *Scolytus multistriatus* (Coleoptera:Scolytidae) to a virgin-female-produced pheromone in the field. *Ann. Entomol. Soc. Am.* **64**, 1143–1149.

476. Peacock, J. W., Silverstein, R. M., Lincoln, A. C., and Simeone, J. B. (1973). Laboratory investigations of the frass of *Scolytus multistriatus* (Coleoptera:Scolytidae) as a source of pheromone. *Environ. Entomol.* **2**, 355–359.

477. Penman, D. R., and Cone, W. W. (1972). Behavior of male twospotted spider mites in response to quiescent female deutonymphs and to web. *Ann. Entomol. Soc. Am.* **65**, 1289–1293.

478. Penman, D. R., and Cone, W. W. (1974). Role of web, tactile stimuli, and female sex pheromone in attraction of male twospotted spider mites to quiescent female deutonymphs. *Ann. Entomol. Soc. Am.* **67**, 179–182.

479. Percy, J. E., and Weatherston, J. (1974). Gland structure and pheromone production in insects. *In* "Pheromones" (M. C. Birch, ed.), pp. 11–34. Am. Elsevier, New York.

480. Persoons, C. J., Minsk, A. K., Voerman, S., Roelofs, W. L., and Ritter, F. J. (1974). Sex pheromones of the moth, *Archips podana*: Isolation, identification and field evaluation of two synergistic geometrical isomers. *J. Insect Physiol.* **20**, 1181–1188.

481. Pfeiffer, W. (1963). Alarm substances. *Experientia* **19**, 1–11.

482. Pfeiffer, W. (1947). Pheromones in fish and amphibia. *In* "Pheromones" (M. C.

Birch, ed.) , pp. 269–296. Am. Elsevier, New York.

483. Piston, J. J., and Lanier, G. N. (1974) . Pheromones of *Ips pini* (Coleoptera: Scolytidae) : Response to interpopulational hybrids and relative attractiveness of males boring in two host species. *Can. Entomol.* **106**, 247–251.

484. Pitman, G. B. (1969) . Pheromone response in pine bark beetles: Influence of host volatiles. *Science* **166**, 905–906.

485. Pitman, G. B., and Vité, J. P. (1963) . Studies on the pheromone of *Ips confusus* (Lec.) . I. Secondary sexual dimorphism in the hindgut epithelium. *Contrib. Boyce Thompson Inst.* **22**, 221–226.

486. Pitman, G. P., and Vité, J. P. (1969) . Aggregation behavior of *Dendroctonus ponderosae* (Coleoptera:Scolytidae) in response to chemical messengers. *Can. Entomol.* **101**, 143–149.

487. Pitman, G. B., and Vité, J. P. (1970) . Field response of *Dendroctonus pseudotsugae* (Coleoptera:Scolytidae) to synthetic frontalin. *Ann. Entomol. Soc. Am.* **63**, 661–664.

488. Pitman, G. B., Kliefoth, R. A., and Vité, J. P. (1965) . Studies on the pheromone of *Ips confusus* (Leconte) . II. Further observations on the site of production. *Contrib. Boyce Thompson Inst.* **23**, 13–18.

489. Pliske, T. E., and Eisner, T. (1969) . Sex pheromone of the queen butterfly: Biology. *Science* **164**, 1170–1172.

490. Poglayen-Neuwall, I. (1966) . On the marking behaviour of the kinkajou (*Potos flavus* Schreber) . *Zoologica (N.Y.)* **51**, 137–141.

491. Priesner, E. (1970) . Ueber die Spezifitaet der Lepidopteren-Sexuallockstoffe und ihre Rolle bei der Artbildung. *Verh. Dtsch. Zool. Ges.* **64**, 337–343.

492. Prokopy, R. J. (1972) . Evidence for a marking pheromone deterring repeated oviposition in apple maggot flies. *Environ. Entomol.* **1**, 326–332.

493. Prokopy, R. J., and Bush, G. L. (1972) . Mating behavior in *Rhagoletis pomonella* (Diptera:Tephritidae) . III. Male aggregation in response to an arrestant. *Can. Entomol.* **104**, 275–283.

494. Rabb, R. L., and Bradley, J. R. (1970) . Marking host eggs by *Telenomus sphingis*. *Ann. Entomol. Soc. Am.* **63**, 1053–1056.

495. Rahn, R. (1968) . Rôle de la plante-hôte sur l'attractivité sexuelle chez *Acrolepia assectella* Zeller (Lep. Plutellidae) . *C. R. Hebd. Seances Acad. Sci., Ser. D* **266**, 2004–2006.

496. Ralls, K. (1971) . Mammalian scent marking. *Science* **171**, 443–449.

497. Ratner, S. C. (1971) . Behavioral characteristics and functions of pheromones of earthworms. *Psychol. Rec.* **21**, 363–371.

498. Rau, P., and Rau, N. (1929) . The sex attraction and rhythmic periodicity in giant saturniid moths. *Trans. Acad. Sci. St. Louis* **26**, 83–221.

499. Read, J. S., Warren, F. L., and Hewitt, P. H. (1968) . Identification of the sex pheromone of the false codling moth (*Argyroploce leucotreta*) . *Chem. Commun.* 1968, 792–793.

500. Redfern, R. E., Cantu, E., Jones, W. A., and Jacobson, M. (1971) . Response of the male southern armyworm in a field cage to Prodenialure A and Prodenialure B. *J. Econ. Entomol.* **64**, 1570–1571.

501. Regnier, F. E., and Wilson, E. O. (1968) . The alarm-defense system of the ant *Acanthomyops claviger*. *J. Insect Physiol.* **14**, 955–970.

502. Renwick, J. A. A. (1967) . Identification of two oxygenated terpenes from the bark beetles *Dendroctonus frontalis* and *Dendroctonus brevicomis*. *Contrib. Boyce Thompson Inst.* **23**, 355–360.

503. Renwick, J. A. A., and Vité, J. P. (1969) . Bark beetle attractants: Mechanism of

colonization by *Dendroctonus frontalis*. *Nature (London)* **224**, 1222–1223.

504. Renwick, J. A. A., and Vité, J. P. (1970). Systems of chemical communication in *Dendroctonus*. *Contrib. Boyce Thompson Inst.* **24**, 283–292.

505. Ressler, R. H., Cialdini, R. B., Ghoca, M. L., and Kleist, S. M. (1968). Alarm pheromone in the earthworm *Lumbricus terrestris*. *Science* **161**, 597–599.

506. Reynierse, J. H., Auld, K. G., and Scavio, M. J., Jr. (1969). Preliminary note on seasonal production of planarian pheromone. *Psychol. Rep.* **24**, 705–706.

507. Reynierse, J. H., Gleason, K. K., and Ottemann, R. (1969). Mechanisms producing aggregations in Planaria. *Anim. Behav.* **17**, 47–63.

508. Ribbands, C. R., and Speirs, N. (1953). The adaptability of the homecoming honeybee. *Br. J. Anim. Behav.* **1**, 59–66.

509. Riddiford, L. M. (1967). *Trans*-2-hexenal: Mating stimulant for polyphemus moths. *Science* **158**, 139–141.

510. Riddiford, L. M., and Williams, C. M. (1967). Volatile principle from oak leaves: Role in sex life of the Polyphemus moth. *Science* **155**, 589–590.

511. Riddiford, L. M., and Williams, C. M. (1967). Chemical signaling between moths and host plant. *Science* **156**, 541.

512. Riddiford, L. M., and Williams, C. M. (1971). Role of the corpora cardiaca in the behavior of saturniid moths. I. Release of sex pheromone. *Biol. Bull.* **140**, 1–7.

513. Riedel, S. M., and Blum, M. S. (1972). Rapid adaptation by paired queens of the honey bee, *Apis mellifera*. *Ann. Entomol. Am.* **65**, 825–829.

514. Riley, R. G., Silverstein, R. M., and Moser, J. C. (1974). Isolation, identification, synthesis and biological activity of volatile compounds from the heads of *Atta* ants. *J. Insect Physiol.* **20**, 1629–1637.

515. Ritter, F. J., Rotgans, I. E. M., Talman, E., Verwiel, P. E. J., and Stein, F. (1973). 5-Methyl-3-butyl-octahydroindolizine, a novel type of pheromone attractive to Pharaoh's ants (*Monomorium pharaonis* (L.)). *Experientia* **29**, 530.

516. Robertson, A., Drage, D. J., and Cohen, M. H. (1972). Control of aggregation in *Dictyostelium discoideum* by an external periodic pulse of cyclic adenosine monophosphate. *Science* **175**, 333–335.

517. Robertson, P. L. (1971). Pheromones involved in aggressive behaviour in the ant, *Myrmecia gulosa*. *J. Insect Physiol.* **17**, 691–715.

518. Rodin, J. O., Silverstein, R. M., Burkholder, W. E., and Gorman, J. E. (1969). Sex attractant of female dermestid beetle *Trogoderma inclusum* LeComte. *Science* **165**, 904–905.

519. Roelofs, W. L., and Arn, H. (1968). Sex attractant of the red-banded leafroller moth. *Nature (London)* **219**, 513.

520. Roelofs, W. L., and Cardé, R. T. (1971). Hydrocarbon sex pheromone in tiger moths (Arctiidae). *Science* **171**, 684–686.

521. Roelofs, W. L., and Cardé, R. T. (1974). Sex pheromones in the reproductive isolation of lepidopterous species. *In* "Pheromones" (M. C. Birch, ed.), pp. 96–114. Am. Elsevier, New York.

522. Roelofs, W. L., and Comeau, A. (1969). Sex pheromone specificity: Taxonomic and evolutionary aspects in Lepidoptera. *Science* **165**, 398–400.

523. Roelofs, W. L., and Comeau, A. (1971). Sex pheromone perception: Synergists and inhibitors for the red-banded leaf roller attractant. *J. Insect Physiol.* **17**, 435–448.

524. Roelofs, W. L., and Comeau, A. (1971). Sex attractants in Lepidoptera. *Pestic. Chem., Proc. Int. IUPAC Congr. Pestic. Chem., 2nd, 1971* Vol. 3, pp. 91–114.

525. Roelofs, W. L., and Tette, J. P. (1970). Sex pheromone of the oblique-banded leaf roller moth. *Nature (London)* **226**, 1172.

526. Roelofs, W. L., Comeau, A., and Selle, R. (1969). Sex pheromone of the oriental fruit moth. *Nature (London)* **224**, 723.

527. Roelofs, W., Comeau, A., Hill, A., and Milicevic, G. (1971). Sex attractant of the codling moth: Characterization with electroantennogram technique. *Science* **174**, 297–299.

528. Roelofs, W. L., Tette, J. P., Taschenberg, E. F., and Comeau, A. (1971). Sex pheromone of the grape berry moth: Identification by classical and electroantennogram methods, and field tests. *J. Insect Physiol.* **17**, 2235–2243.

529. Roelofs, W. L., Cardé, R. T., Bartell, R. J., and Tierney, P. G. (1972). Sex attractant trapping of the European corn borer in New York. *Environ. Entomol.* **1**, 606–608.

530. Roelofs, W. L., Hill, A. S., Cardé, R. T., Tette, J. P., Madsen, H., and Vakenti, J. (1974). Sex pheromones of the fruit tree leafroller moth, *Archips argyrospilus*. *Environ. Entomol.* **3**, 747–751.

531. Roelofs, W. L., Hill, A. S., Cardé, R. T., and Baker, T. C. (1974). Two sex pheromone components of the tobacco budworm moth, *Heliothis virescens*. *Life Sci.* **14**, 1555–1562.

532. Roelofs, W. L., Hill, A. S., and Cardé, R. T. (1975). Sex pheromone components of the redbanded leafroller, *Argyrotaenia velutinana* (Lepidoptera:Tortricidae). *J. Chem. Ecol.* **1**, 83–89.

533. Rogoff, W. M., Beltz, A. D., Johnsen, J. O., and Plapp, F. W. (1964). A sex pheromone in the housefly, *Musca domestica* L. *J. Insect Physiol.* **10**, 239–246.

534. Röller, H., Biemann, K., Bjerke, J. S., Norgard, D. W., and McShan, W. H. (1968). Sex pheromones of pyralid moths. I. Isolation and identification of the sex-attractant of *Galleria mellonella* L. (greater waxmoth). *Acta Entomol. Bohemoslov.* **65**, 208–211.

535. Ropartz, P. (1966). Mise en évidence d'une odeur de groupe chez les souris par la mesure de l'activité locomotrice. *C. R. Hebd. Seances Acad. Sci., Ser. D* **262**, 507–510.

536. Ropartz, P. (1966). Mise en évidence du rôle d'une secretion odorante des glands sudoripares dans la regulation de l'activité locomotive chez la souris. *C. R. Hebd. Seances Acad. Sci., Ser. D* **263**, 525–528.

537. Ropartz, P. (1968). Rôle des communications olfactives dans le comportement social des souris males. *Colloq. Int. C.N.R.S.* **173**, 323–339.

538. Ropartz, P. (1968). The relation between olfactory stimulation and aggressive behaviour in mice. *Anim. Behav.* **16**, 97–100.

539. Roth, L. M., and Cohen, S. (1973). Aggregation in Blattaria. *Ann. Entomol. Soc. Am.* **66**, 1315–1323.

540. Roth, L. M., and Dateo, G. P. (1966). A sex pheromone produced by males of the cockroach *Nauphoeta cinerea*. *J. Insect Physiol.* **12**, 255–265.

541. Roth, L. M., and Willis, E. R. (1952). A study of cockroach behavior. *Am. Midl. Nat.* **47**, 66–129.

542. Rothschild, G. H. L., and Minks, A. K. (1974). Time of activity of male oriental fruit moths at pheromone sources in the field. *Environ. Entomol.* **3**, 1003–1007.

543. Rottman, S. J., and Snowdon, C. T. (1972). Demonstration and analysis of an alarm pheromone in mice. *J. Comp. Physiol. Psychol.* **81**, 483–490.

544. Rudinsky, J. A. (1963). Response of *Dendroctonus pseudotsugae* Hopk. to volatile attractants. *Contrib. Boyce Thompson Inst.* **22**, 23–38.

545. Rudinsky, J. A. (1969). Masking of the aggregation pheromone in *Dendroctonus pseudotsugae* Hopk. *Science* **166**, 884–885.

546. Rudinsky, J. A. (1973). Multiple functions of the Douglas fir beetle pheromone 3-methyl-2-cyclohexen-1-one. *Environ. Entomol.* **2**, 579–585.

547. Rudinsky, J. A., and Daterman, G. E. (1964). Response of the ambrosia beetle *Trypodendron lineatum* (Oliv.) to a female-produced pheromone. *Z. Angew. Entomol.* **54**, 300–303.

548. Rudinsky, J. A., and Daterman, G. E. (1964). Field studies on flight patterns and olfactory responses of ambrosia beetles in Douglas-fir forests of western Oregon. *Can. Entomol.* **96**, 1339–1352.

549. Rudinsky, J. A., and Michael, R. R. (1972). Sound production in Scolytidae: Chemostimulus of sonic signal by the Douglas-fir beetle. *Science* **751**, 1386–1390.

550. Rudinsky, J. A., and Michael, R. R. (1974). Sound production in Scolytidae: 'Rivalry' behaviour of male *Dendroctonus* beetles. *J. Insect Physiol.* **20**, 1219–1230.

551. Rudinsky, J. A., Novák, V., and Svihra, P. (1971). Attraction of the bark beetle *Ips typographus* L. to terpenes and a male-produced pheromone. *Z. Angew. Entomol.* **67**, 179–188.

552. Rudinsky, J. A., Furniss, M. M., Kline, L. N., and Schmitz, R. F. (1972). Attraction and repression of *Dendroctonus pseudotsugae* (Coleoptera:Scolytidae) by three synthetic pheromones in traps in Oregon and Idaho. *Can. Entomol.* **104**, 815–822.

553. Rudinsky, J. A., Kinzer, G. W., Fentiman, A. F., Jr., and Foltz, R. L. (1972). *trans*-Verbenol isolated from Douglas-fir beetle: Laboratory and field bioassays in Oregon. *Environ. Entomol.* **1**, 485–488.

554. Rudinsky, J. A., Morgan, M., Libbey, L. M., and Michael, R. R. (1973). Sound production in Scolytidae: 3-methyl-2-cyclohexen-1-one released by the female Douglas fir beetle in response to male sonic signal. *Environ. Entomol.* **2**, 505–509.

555. Rudinsky, J. A., Morgan, M. E., Libbey, L. M., and Putnam, T. B. (1974). Anti-aggregative-rivalry pheromone of the mountain pine beetle and a new arrestant of the southern pine beetle. *Environ. Entomol.* **3**, 90–98.

556. Rudinsky, J. A., Morgan, M. E., Libbey, L. M., and Putnam, T. B. (1974). Additional components of the Douglas fir beetle (Col., Scolytidae) aggregative pheromone and their possible utility in pest control. *Z. Angew. Entomol.* **76**, 65–77.

557. Ryan, E. P. (1966). Pheromone: Evidence in a decapod crustacean. *Science* **151**, 340–341.

558. Saario, C. A., Shorey, H. H., and Gaston, L. K. (1970). Sex pheromones of noctuid moths. XIX. Effect of environmental and seasonal factors on captures of males of *Trichoplusia ni* in pheromone-baited traps. *Ann. Entomol. Soc. Am.* **63**, 667–672.

559. Salm, R. W., and Fried, B. (1973). Heterosexual chemical attraction in *Camallanus* Sp. (Nematoda) in the absence of worm-mediated tactile behavior. *J. Parasitol.* **59**, 434–436.

560. Salt, G. (1937). The sense used by *Trichogramma* to distinguish between parasitized hosts. *Proc. R. Soc. London, Ser. B* **122**, 57–75.

561. Sanders, C. J. (1969). Extrusion of the female sex pheromone gland in the eastern spruce budworm, *Choristoneura fumiferana* (Lepidoptera:Tortricidae). *Can. Entomol.* **101**, 760–762.

562. Sanders, C. J. (1971). Sex pheromone specificity and taxonomy of budworm moths *(Choristoneura)*. *Science* **171**, 911–913.

563. Sanders, C. J., and Lucuik, G. S. (1972). Factors affecting calling by female eastern spruce budworm, *Choristoneura fumiferana* (Lepidoptera:Tortricidae). *Can. Entomol.* **104**, 1751–1762.

564. Sarmiento, R., Beroza, M., Bierl, B. A., and Tardif, J. G. R. (1972). Activity of compounds related to disparlure, the sex attractant of the gypsy moth. *J. Econ. Entomol.* **65**, 665–667.

565. Schein, M. W., and Hale, E. B. (1965). Stimuli eliciting sexual behaviour. *In* "Sex and Behavior" (F. A. Beach, ed.), pp. 440–482. Wiley, New York.

566. Schloeth, R. (1956). Zur Psychologie der Begegnung zwischen Tieren. *Bahaviour* **10,** 1–79.

567. Schneider, D. (1962). Electrophysiological investigation on the olfactory specificity of sexual attracting substances in different species of moths. *J. Insect Physiol.* **8,** 15–30.

568. Schneider, D. (1964). Insect antennae. *Annu. Rev. Entomol.* **9,** 193–122.

569. Schneider, D. (1967). Wie arbeitet der Geruchssinn bei Mensch und Tier? *Naturwissenschaften* **20,** 319–326.

570. Schneider, D. (1969). Insect olfaction: Deciphering system for chemical messages. *Science* **163,** 1031–1037.

571. Schneider, D. (1970). Olfactory receptors for the sexual attractant (bombykol) of the silk moth. *In* "The Neurosciences: Second Study Program" (F. O. Schmitt, ed.), pp. 511–518. Rockefeller Univ. Press, New York.

572. Schneider, D., Lange, R., Schwarz, F., Beroza, M., and Bierl, B. A.(1974). Attraction of male gypsy and nun moths to disparlure and some of its chemical analogues. *Oecologia* **14,** 19–36.

573. Schneirla, T. C. (1933). Studies on army ants in Panama. *J. Comp. Psychol.* **15,** 267–299.

574. Schneirla, T. C., and Brown, R. Z. (1950). Army-ant life and behavior under dry-season conditions. 4. Further investigation of cyclic processes in behavioral and reproductive functions. *Bull. Am. Mus. Nat. Hist.* **95,** 263–354.

575. Schneirla, T. C., and Piel, G. (1948). The army ant. *Sci. Am.* **178,** 16–23.

576. Schönherr, J. (1972). Pheromon beim Kiefern-Borkenkaefer "Waldgaertuer", *Myelophilus piniperda* L. (Coleopt., Scolytidae). *Z. Angew. Entomol.* **71,** 410–413.

577. Schultze-Westrum, T. (1965). Innerartliche Verständigung durch Düfte beim Gleitbeutler, *Petaurus breviceps papuanus* Thomas (Marsupialia, Phalangeridae). *Z. Vergl. Physiol.* **50,** 151–220.

578. Schutz, F. (1956). Vergleichende Untersuchungen über die Schreckreaktion bei Fischen und deren Verbreitung. *Z. Vergl. Physiol.* **38,** 84–135.

579. Schwinck, I. (1954). Experimentelle Untersuchungen über Geruchseiun und Strömungswahrnehmung in der Orientierung bei Nachtschmetterlingen. *Z. Vergl. Physiol.* **37,** 19–56.

580. Schwinck, I. (1955). Freilandversuche zur Frage der Artspezifität des weiblichen Sexualduftstoffes der Nonne (*Lymantria monacha* L.) und des Schwammspinsners (*Lymantria dispar* L.). *Z. Angew. Entomol.* **37,** 349–357.

581. Schwinck, I. (1955). Weitere untersuchungen zur Frage der Geruchsorientierung der Nachtschmetterlinge: Partielle Fühleramputation bei Spinnermännchen, Insbesondere am Seidenspinner *Bombyx mori* L. *Z. Vergl. Physiol.* **37,** 439–458.

582. Schwinck, I. (1958). A study of olfactory stimuli in the orientation of moths. *Proc. Int. Congr. Entomol., 10th, 1956* Vol. 2, pp. 577–582.

583. Seitz, E. (1969). Die Bedeutung geruchlicher Orientierung beim Plumplori *Nycticebus coucang* Boddaert 1758 (Prosimii, Lorisidae). *Z. Tierpsychol.* **26,** 73–103.

584. Sekul, A. A., and Sparks, A. N. (1967). Sex pheromone of the fall armyworm moth: Isolation, identification, and synthesis. *J. Econ. Entomol.* **60,** 1270–1272.

585. Shaposhnikova, N. G., and Gavrilov, B. N. (1973). O vospriyatii rabochimi pchelami (*Apis mellifera* Caucasica) sinetichekogo feromona matochnogo veshchestra *trans*-9-ketodetsen-2-oboi kisloty. *Zool. Zh.* **52,** 291–293.

586. Sharma, R. K., Rice, R. E., Reynolds, H. T., and Shorey, H. H. (1971). Seasonal influence and effect of trap location on catches of pink bollworm males in sticky traps baited with hexalure. *Ann. Entomol. Soc. Am.* **64,** 102–105.

587. Shearer, D. A., and Boch, R. (1965). 2-Heptanone in the mandibular gland secre-

tion of the honey bee. *Nature (London)* **206**, 530.

588. Shearer, D. A., and Boch, R. (1966). Citral in the Nassanoff pheromone of the honey bee. *J. Insect Physiol.* **12**, 1513–1521.

589. Shorey, H. H. (1964). Sex pheromones of noctuid moths. II. Mating behavior of *Trichoplusia ni* (Lepidoptera:Noctuidae) with special reference to the role of the sex pheromone. *Ann. Entomol. Soc. Am.* **57**, 371–377.

590. Shorey, H. H. (1966). The biology of *Trichoplusia ni* (Lepidoptera:Noctuidae). IV. Environmental control of mating. *Ann. Entomol. Soc. Am.* **59**, 502–506.

591. Shorey, H. H. (1970). Sex pheromones of Lepidoptera. *In* "Control of Insect Behavior by Natural Products" (D. L. Wood, R. M. Silverstein, and M. Nakajima, eds.), pp. 249–284. Academic Press, New York.

592. Shorey, H. H. (1973). Behavioral responses to insect pheromones. *Annu. Rev. Entomol.* **18**, 349–380.

593. Shorey, H. H. (1974). Environmental and physiological control of insect sex pheromone behavior. *In* "Pheromones" (M. C. Birch, ed.), pp. 62–80. Am. Elsevier, New York.

594. Shorey, H. H., and Farkas, S. R. (1973). Sex pheromones of Lepidoptera. 42. Terrestrial odor-trail following by pheromone-stimulated males of *Trichoplusia ni*. *Ann. Entomol. Soc. Am.* **66**, 1213–1214.

595. Shorey, H. H., and Gaston, L. K. (1964). Sex pheromones of noctuid moths. III. Inhibition of male responses to the sex pheromone in *Trichoplusia ni* (Lepidoptera:Noctuidae). *Ann. Entomol. Soc. Am.* **57**, 775–779.

596. Shorey, H. H., and Gaston, L. K. (1965). Sex pheromones of noctuid moths. V. Circadian rhythm of pheromone-responsiveness in males of *Autographa californica, Heliothis virescens, Spodoptera exigua,* and *Trichoplusia ni* (Lepidoptera:Noctuidae). *Ann. Entomol. Soc. Am.* **58**, 597–600.

597. Shorey, H. H., and Gaston L. K. (1965). Sex pheromones of noctuid moths. VII. Quantitative aspects of the production and release of pheromone by females of *Trichoplusia ni* (Lepidoptera:Noctuidae). *Ann. Entomol. Soc. Am.* **58**, 604–608.

598. Shorey, H. H., and Gaston, L. K. (1970). Sex pheromones of noctuid moths. XX. Short-range visual orientation by pheromone-stimulated males of *Trichoplusia ni*. *Ann. Entomol. Soc. Am.* **63**, 829–832.

599. Shorey, H. H., Gaston, L. K., and Roberts, J. S. (1965). Sex pheromones of noctuid moths. VI. Absence of behavioral specificity for the female sex pheromones of *Trichoplusia ni* versus *Autographa californica,* and *Heliothis zea* versus *Heliothis virescens* (Lepidoptera:Noctuidae). *Ann. Entomol. Soc. Am.* **58**, 600–603.

600. Shorey, H. H., McFarland, S. U., and Gaston, L. K. (1968). Sex pheromones of noctuid moths. XIII. Changes in pheromone quantity, as related to reproductive age and mating history, in females of seven species of Noctuidae (Lepidoptera). *Ann. Entomol. Soc. Am.* **61**, 372–376.

601. Shorey, H. H., Morin, K. L., and Gaston, L. K. (1968). Sex pheromones of noctuid moths. XV. Timing of development of pheromone-responsiveness and other indicators of reproductive age in males of eight species. *Ann. Entomol. Soc. Am.* **61**, 857–861.

602. Shorey, H. H., Gaston, L. K., and Jefferson, R. N. (1968). Insect sex pheromones. *Adv. Pest Control Res.* **8**, 57–126.

603. Shorey, H. H., Bartell, R. J., and Barton Browne, L. B. (1969). Sexual stimulation of males of *Lucilia cuprina* (Calliphoridae) and *Drosophila melanogaster* (Drosophilidae) by the odors of aggregation sites. *Ann. Entomol. Soc. Am.* **62**, 1419–1421.

604. Signoret, J. P., and du Mesnil du Buisson, F. (1962). Étude du comportement de la

truie en oestrus. *Proc. Int. Congr. Anim. Reprod., 4th, 1961* Vol. 2, pp. 171–175.

605. Silverstein, R. M., Rodin, J. O., and Wood, D. L. (1966). Sex attractants in frass produced by male *Ips confusus* in ponderosa pine. *Science* **154**, 509–510.

606. Silverstein, R. M., Rodin, J. O., Burkholder, W. E., and Gorman, J. O. (1967). Sex attractant of the black carpet beetle. *Science* **157**, 85–87.

607. Silverstein, R. M., Brownlee, R. G., Bellas, T. E., Wood, D. L., and Browne, L. E. (1968). Brevicomin: Principal sex attractant in the frass of the female western pine beetle. *Science* **159**, 889–891.

608. Simpson, J. (1963). Queen perception by honey bee swarms. *Nature (London)* **199**, 94–95.

609. Sink, J. D. (1967). Theoretical aspects of sex odor in swine. *J. Theor. Biol.* **17**, 174–180.

610. Smith, R. G., Daterman, G. E., Daves, G. D., Jr., McMurtrey, K. D., and Roelofs, W. L. (1974). Sex pheromone of the European pine shoot moth: Chemical identification and field tests. *J. Insect Physiol.* **20**, 661–668.

611. Smythe, R. V., and Coppel, H. C. (1966). A preliminary study of the sternal gland of *Reticulitermes flavipes* (Isoptera:Rhinotermitidae). *Ann. Entomol. Soc. Am.* **59**, 1008–1010.

612. Smythe, R. V., Coppel, H. C., Lipton, S. H., and Strong, F. M. (1967). Chemical studies of attractants associated with *Reticulitermes flavipes* and *R. virginicus*. *J. Econ. Entomol.* **60**, 228–233.

613. Snyder, N. F. R. (1967). An alarm reaction of aquatic gastropods to intraspecific extract. *N.Y., Agr. Exp. Stn., Ithaca, Mem.* **403**, 1–122.

614. Snyder, N., and Snyder, H. (1970). Alarm response of *Diadema antillarum*. *Science* **168**, 276–278.

615. Snyder, N. F. R., and Snyder, H. A. (1971). Pheromone-mediated behaviour of *Fasciolaria tulipa*. *Anim. Behav.* **19**, 257–268.

616. Somers, J.-A., Shorey, H. H., and Gaston, L. K. (1976). Conditions affecting sex pheromone response in *Rhabditis pellio*. *J. Chem. Ecol.* (submitted for publication).

617. Soo Hoo, C. F., and Roberts, R. J. (1965). Sex attraction in *Rhopaea* (Coleoptrea: Scarabaeidae). *Nature (London)* **205**, 724–725.

618. Sower, L. L., Shorey, H. H., and Gaston, L. K. (1970). Sex pheromones of noctuid moths. XXI. Light:dark cycle regulation and light inhibition of sex pheromone release by females of *Trichoplusia ni*. *Ann. Entomol. Soc. Am.* **63**, 1090–1092.

619. Sower, L. L., Gaston, L. K., and Shorey, H. H. (1971). Sex pheromones of noctuid moths. XXVI. Female release rate, male response threshold, and communication distance for *Trichoplusia ni*. *Ann. Entomol. Soc. Am.* **64**, 1448–1456.

620. Sower, L. L., Shorey, H. H., and Gaston, L. K. (1971). Sex pheromones of noctuid moths. XXV. Effects of temperature and photoperiod on circadian rhythms of sex pheromone release by females of *Trichoplusia ni*. *Ann. Entomol. Soc. Am.* **64**, 448–492.

621. Sower, L. L., Long, J. S., Vick, K. W., and Coffelt, J. A. (1973). Sex pheromone of the angoumois grain moth: Effects of habituation on the pheromone response of the male. *Ann. Entomol. Soc. Am.* **66**, 991–995.

622. Sower, L. L., Kaae, R. S., and Shorey, H. H. (1973). Sex pheromones of Lepidoptera XLI. Factors limiting potential distance of sex pheromone communication in *Trichoplusia ni*. *Ann. Entomol. Soc. Am.* **66**, 1121–1122.

623. Sower, L. L., Vick, K. W., and Bull, K. A. (1974). Perception of olfactory stimuli that inhibit the responses of male phycitid moths to sex pheromones. *Environ. Entomol.* **3**, 277–279.

624. Sower, L. L., Vick, K. W., and Tumlinson, J. H. (1974). (Z,E)-9,12-Tetradecadien-l-ol: A chemical released by female *Plodia interpunctella* that inhibits the sex pheromone response of male *Cadra cautella*. *Environ. Entomol.* **3**, 120–122.

625. Stein, G. (1963). Untersuchungen über den Sexuallockstoff der Hummelmännchen. *Biol. Zentralbl.* **82**, 343–349.

626. Stein, G. (1963). Über den Sexuallockstoff von Hummelmännchen. *Naturwissenschaften* **50**, 305.

627. Steinbrecht, R. A. (1964). Die Abhängigkeit der Lockwirkung des Sexualduftorgans weiblicher Seidenspinner (*Bombyx mori*) von Alter und Kopulation. *Z. Vergl. Physiol.* **48**, 341–356.

628. Steinbrecht, R. A. (1969). Comparative morphology of olfactory receptors. *In* "Olfaction and Taste" (C. Pfaffmann, ed.), pp. 3–21. Rockefeller Univ. Press, New York.

629. Steinbrecht, R. A. (1970). Zur morphometrie der antenne des seidenspinners, *Bombyx mori* L.: Zahl und verteilung der riechsensillen (Insecta, Lepidoptera). *Z. Morphol. Tiere* **68**, 93–126.

630. Stinson, C. G., and Patterson, R. L. S. (1972). C_{19}-Δ^{16} steroids in boar sweat glands. *Br. Vet. J.* **128**, 41–43.

631. Stobbe, R. (1912). Die abdominalen Duftorgane der männlichen Sphingiden und Noctuiden. *Zool. Jahrb.* **32**, 493–532.

632. Stringfellow, F. (1974). Hydroxyl ion, an attractant to the male of *Pelodera strongyloides*. *Proc. Helminthol. Soc. Wash.* **41**, 4–10.

633. Stuart, A. M. (1961). Mechanisms of trail-laying in two species of termites. *Nature (London)* **189**, 419.

634. Stuart, A. M. (1963). The origin of the trail in the termites *Nasutitermes corniger* (Motschulsky) and *Zootermopsis nevadensis* (Hagen), Isoptera. *Physiol. Zool.* **36**, 69–84.

635. Stuart, A. M. (1963). Studies on the communications of alarm in the termite *Zootermopsis nevadensis* (Hagen), Isoptera. *Physiol. Zool.* **36**, 85–96.

636. Stuart, A. M. (1967). Alarm, defense, and construction behavior relationships in termites (Isoptera). *Science* **156**, 1123–1125.

637. Stuart, A. M. (1970). The role of chemicals in termite communication. *Adv. Chemoreception* **1**, 79–106.

638. Stumper, R. (1956). Sur les sécrétions attractives des fourmis femelles. *C. R. Hebd. Seances Acad. Sci.* **242**, 2487–2489.

639. Su, H. C. F., and Mahany, P. G. (1974). Synthesis of the sex pheromone of the female angoumois grain moth and its geometric isomers. *J. Econ. Entomol.* **67**, 319–321.

640. Sutton, O. G. (1953). "Micrometeorology. A Study of Physical Processes in the Lowest Layers of the Earth's Atmosphere." McGraw-Hill, New York.

641. Szabo, T. I. (1974). Behavioural studies on queen introduction in honey bees. 4. Introduction of queen models into honey bee colonies. *Am. Bee. J.* **114**, 174–176.

642. Takahashi, F. (1973). Sex pheromones; are they really species specific? *Contrib. Entomol. Lab. Coll. Agr., Kyoto Univ.* No. 444, pp. 13–21.

643. Takahashi, F. T., and Kittredge, J. S. (1973). Suppression of the feeding response in the crab *Pachygrapsus crassipes:* Pheromone induction. *Tex. Rep. Biol. Med.* **31**, 403–408.

644. Talbot, L. M., and Talbot, M. H. (1963). The wildebeest in western Masailand, East Africa. *Wildl. Monogr.* **12**, 1–88.

645. Tamaki, Y., and Yushima, T. (1974). Biological activity of the synthesized sex pheromone and its geometrical isomers of *Spodoptera litura* (F.) (Lepidoptera:

Noctuidae) . *Appl. Entomol. Zool.* **9**, 73–79.

646. Tamaki, Y., and Yushima, T. (1974) . Sex pheromone of the cotton leafworm, *Spodoptera littoralis. J. Insect Physiol.* **20**, 1005–1014.

647. Tamaki, Y., Noguchi, H., and Yushima, T. (1971) . Two sex pheromones of the smaller tea tortrix: Isolation, identification, and synthesis. *Appl. Entomol. Zool.* **6**, 139–141.

648. Tamaki, Y., Noguchi, H., Yushima, T., Hirano, C., Honma, K., and Sugawara, H. (1971) . Sex pheromone of the summerfruit tortrix: Isolation and identification. *Konchu* **39**, 338–340.

649. Tamaki, Y., Noguchi, H., and Yushima, T. (1973) . Sex pheromone of *Spodoptera litura* (F.) (Lepidoptera:Noctuidae) : Isolation, identification, and synthesis. *Appl. Entomol. Zool.* **8**, 200–203.

650. Tarasov, P. P. (1960) . Biological significance of scent glands in mammals. *Zool. Zh.* **39**, 1062–1068 (in Russian) .

651. Tavolga, W. N. (1956) . Visual, chemical, and sound stimuli as cues in the sex discriminatory behavior of the gobiid fish *Bathygobius soporator. Zoologica (N.Y.)* **41**, 49–64.

652. Teale, E. W. (1961). "The Insect World of J. Henri Fabre." Dodd, Mead, New York.

653. Teetes, G. L., and Randolph, N. M. (1970) . Color preference and sex attraction among sunflower moths. *J. Econ. Entomol.* **63**, 1358–1359.

654. Tette, J. P. (1974) . Pheromones in insect population management. *In* "Pheromones" (M. C. Birch, ed.) , pp. 399–410. Am. Elsevier, New York.

655. Thiessen, D. D., Friend, H. C., and Lindzey, G. (1968) . Androgen control of territorial marking in the Mongolian gerbil *(Meriones unguiculatus)* . *Science* **160**, 432–434.

656. Thiessen, D. D., Blum, S. L., and Lindzey, G. (1970) . A scent marking response associated with the ventral sebaceous gland of the mongolian gerbil *(Meriones unguiculatus)* . *Anim. Behav.* **18**, 26–30.

657. Thiessen, D. D., Lindzey, G., Blum, S. L., and Wallace, P. (1971) . Social interactions and scent marking in the Mongolian gerbil *(Meriones unguiculatus)* . *Anim. Behav.* **19**, 505–513.

658. Thiessen, D. D., Regnier, F. E., Rice, M., Goodwin, M., Isaacks, N., and Lawson, N. (1974) . Identification of a ventral scent marking pheromone in the male Mongolian gerbil *(Meriones unguiculatus)* . *Science* **184**, 83–85.

659. Thines, G., and Vandenbussche, E. (1966) . The effects of alarm substance on the schooling behaviour of *Rasbora heteromorpha* Duncker in day and night conditions. *Anim. Behav.* **14**, 296–302.

660. Thomson, J. A., and Pears, F. N. (1962) . The functions of the anal glands of the brushtail possum. *Victorian Nat., Melbourne* **78**, 306–308.

661. Timms, A. M., and Kleerekoper, H. (1972) . The locomotor responses of male *Ictalurus punctatus*, the channel catfish, to a pheromone released by the ripe female of the species. *Trans. Am. Fish Soc.* **101**, 302–310.

662. Toba, H. H., Kishaba, A. N., and Wolf, W. W. (1968) . Bioassay of the synthetic female sex pheromone of the cabbage looper. *J. Econ. Entomol.* **61**, 812–816.

663. Todd, J. H., Atema, J., and Bardach, J. E. (1967) . Chemical communication in the social behavior of a fish, the yellow bullhead, *Icatalurus natalis. Science* **158**, 672–673.

664. Torgerson, R. L., and Akre, R. D. (1970) . The persistence of army ant chemical trails and their significance in the ecitonine-ecitophile association (Formicidae: Ecitonini) . *Melanderia* **5**, 1–28.

665. Traynier, R. M. M. (1968) . Sex attraction in the Mediterranean flour moth, *Ana-*

gasta kühniella: Location of the female by the male. *Can. Entomol.* **100**, 5–10.

666. Traynier, R. M. M. (1970). Habituation of the response to sex pheromone in two species of Lepidoptera, with reference to a method of control. *Entomol. Exp. Appl.* **13**, 179–187.

667. Traynier, R. M. M. (1970). Sexual behaviour of the Mediterranean flour moth, *Anagasta kühniella:* Some influence of age, photoperiod, and light intensity. *Can. Entomol.* **102**, 534–540.

668. Tricot, M.-C., Pasteels, J. M., and Tursch, B. (1972). Phéromones stimulant et inhibant l'aggressivité chez *Myrmica rubra. J. Insect Physiol.* **18**, 499–509.

669. Truman, J. W. (1973). How moths "turn on": A study of the action of hormones on the nervous system. *Am. Sci.* **61**, 700–706.

670. Tumlinson, J. H., Hardee, D. D., Gueldner, R. C., Thompson, A. C., Hedin, P. A., and Minyard, J. P. (1969). Sex pheromones produced by male boll weevil: Isolation, identification, and synthesis. *Science* **166**, 1010–1012.

671. Tumlinson, J. H., Moser, J. C., Silverstein, R. M., Brownlee, R. G., and Ruth, J. M. (1972). A volatile trail pheromone of the leaf-cutting ant, *Atta texana. J. Insect Physiol.* **18**, 809–814.

672. Tumlinson, J. H., Mitchell, E. R., Browner, S. M., and Lindquist, D. A. (1972). A sex pheromone for the soybean looper. *Environ. Entomol.* **1**, 466–468.

673. Tumlinson, J. H., Mitchell, E. R., Browner, S. M., Mayer, M. S., Green, N., Hines, R., and Lindquist, D. A. (1972). *cis*-7-Dodecen-1-ol, a potent inhibitor of the cabbage looper sex pheromone. *Environ. Entomol.* **1**, 354–358.

674. Tumlinson, J. H., Yonce, C. E., Doolittle, R. E., Heath, R. R., Gentry, C. R., and Mitchell, E. R. (1974). Sex pheromones and reproductive isolation of the lesser peachtree borer and the peachtree borer. *Science* **185**, 614–616.

675. Velthuis, H. H. W., and van Es, J. (1964). Some functional aspects of the mandibular glands of the queen honeybee. *J. Apic. Res.* **3**, 11–16.

676. Verron, H. (1963). Rôles des stimuli chimiques dans l'attraction sociale chez *Calotermes flavicollis* (Fabr.). *Insectes Soc.* **10**, 167–336.

677. Vick, K. W., Drummond, P. C., and Coffelt, J. A. (1973). *Trogoderma inclusum* and *T. glabrum:* Effects of time of day on production of female pheromone, male responsiveness, and mating. *Ann. Entomol. Soc. Am.* **66**, 1001–1004.

678. Vick, K. W., Drummond, P. C., Sower, L. L., and Coffelt, J. A. (1973). Sex pheromone habituation: The effects of habituation on the pheromone response level of *Trogoderma inclusum* (Coleoptera:Dermestidae). *Ann. Entomol. Soc. Am.* **66**, 667–670.

679. Vick, K. W., Su, H. C. F., Sower, L. L., Mahaney, P. G., and Drummond, P. C. (1974). (*Z-E*)-7,11-Hexadecadien-1-ol acetate: The sex pheromone of the angoumois grain moth, *Sitotroga cerealella. Experientia* **30**, 17–18.

680. Vinson, S. B. (1972). Courtship behavior and evidence for a sex pheromone in the parasitoid *Campoletis sonorensis* (Hymenoptera:Ichneumonidae). *Environ. Entomol.* **1**, 409–414.

681. Vité, J. P., and Gara, R. I. (1962). Volatile attractants from ponderosa pine attacked by bark beetles (Coleoptera:Scolytidae). *Contrib. Boyce Thompson Inst.* **21**, 251–274.

682. Vité, J. P., and Pitman, G. B. (1967). Concepts in research on bark beetle attraction and manipulation. *Proc. Int. Union Forest. Res. Org. 14th,* **24**, 683–701.

683. Vité, J. P., and Pitman, G. B. (1968). Bark beetle aggregation: Effects of feeding on the release of pheromones in *Dendroctonus* and *Ips. Nature (London)* **218**, 169–170.

684. Vité, J. P., and Pitman, G. B. (1968). Insect and host odors in the aggregation of

the western pine beetle. *Can. Entomol.* **101,** 113–117.

685. Vité, J. P., and Pitman, G. B. (1969). Aggregation behaviour of *Dendroctonus brevicomis* in response to synthetic pheromones. *J. Insect Physiol.* **15,** 1617–1622.

686. Vité, J. P., and Renwick, J. A. A. (1968). Insect and host factors in the aggregation of the southern pine beetle. *Contrib. Boyce Thompson Inst.* **24,** 61–64.

687. Vité, J. P., and Renwick, J. A. A. (1971). Inhibition of *Dendroctonus frontalis* response to frontalin by isomers of brevicomin. *Naturwissenschaften* **58,** 418.

688. Vité, J. P., Gara, R. I., and Kliefoth, R. A. (1963). Collection and bioassay of a volatile fraction attractive to *Ips confusus* (Lec.) (Coleoptera:Scolytidae). *Contrib. Boyce Thompson Inst.* **22,** 39–50.

689. Vité, J. P., Gara, R. I., and von Scheller, H. D. (1964). Field observations on the response to attractants of bark beetles infesting southern pines. *Conrtib. Boyce Thompson Inst.* **22,** 461–470.

690. Vité, J. P., Pitman, G. B., Fentiman, A. F., Jr., and Kinzer, G. W. (1972). 3-Methyl-2-cyclohexen-1-ol isolated from *Dendroctonus.* *Naturwissenschaften* **59,** 469.

691. Vité, J. P., Islas, S. F., Renwick, J. A. A., Hughes, P. R., and Kliefoth, R. A. (1974). Biochemical and biological variation of southern pine beetle populations in North and Central America. *Z. Angew. Entomol.* **75,** 422–435.

692. von Frisch, K. (1938). Zur Physiologie des Fischschwarmes. *Naturwissenschaften* **26,** 601–606.

693. von Frisch, K. (1941). Über einen Schreckstoff der Fischhaut und seine biologische Bedeutung. *Z. Vergl. Physiol.* **29,** 46–145.

694. Waldorf, E. S. (1974). Sex pheromone in the springtail, *Sinella curviseta*. *Environ. Entomol.* **3,** 916–918.

695. Wallis, D. I. (1962). Aggressive behaviour in the ant, *Formica fusca*. *Anim. Behav.* **10,** 267–274.

696. Wallis, D. I. (1963). A comparison of the response to aggressive behaviour in two species of ants, *Formica fusca* and *Formica sanguinea*. *Anim. Behav.* **11,** 164–171.

697. Walsh, J. P., and Tschinkel, W. R. (1974). Brood recognition by contact pheromone in the red imported fire ant, *Solenopsis invicta*. *Anim. Behav.* **22,** 695–704.

698. Watkins, J. F., and Cole, T. W. (1966). The attraction of army ant workers to secretions of their queens. *Tex. J. Sci.* **18,** 254–265.

699. Weatherston, J., Roelofs, W., Comeau, A., and Sanders, C. J. (1971). Studies of physiologically active arthropod secretions. X. Sex pheromone of the eastern spruce budworm, *Choristoneura fumiferana* (Lepidoptera:Tortricidae). *Can. Entomol.* **103,** 1741–1747.

700. Wells, M. J., and Buckley, S. K. L. (1972). Snails and trails. *Anim. Behav.* **20,** 345–355.

701. Wharton, D. R. A., Miller, G. L., and Wharton, M. L. (1954). The odorous attractant of the American cockroach, *Periplaneta americana* (L.). *J. Gen. Physiol.* **37,** 461–469.

702. Wharton, D. R. A., Miller, G. L., and Wharton, M. L. (1954). The odorous attractant of the American cockroach, *Periplaneta americana* (L.). II. A bioassay method for the attractant. *J. Gen. Physiol.* **37,** 471–481.

703. Wharton, M. L., and Wharton, D. R. A. (1957). The production of sex attractant substance and of oothecae by the normal and irradiated American cockroach, *Periplaneta americana* L. *J. Insect Physiol.* **1,** 229–239.

704. Whitten, W. K. (1966). Pheromones and mammalian reproduction. *Adv. Reprod. Physiol.* **1,** 155–177.

705. Wilson, E. O. (1962). Chemical communication among workers of the fire ant

Solenopsis saevissima. 1. The organization of mass-foraging. *Anim. Behav.* **10**, 134–147.

706. Wilson, E. O. (1963). Pheromones. *Sci. Am.* **208**, 100–114.
707. Wilson, E. O. (1963). The social biology of ants. *Annu. Rev. Entomol.* **8**, 345–368.
708. Wilson, E. O. (1965). Chemical communication in the social insects. *Science* **149**, 1064–1071.
709. Wilson, E. O. (1968). Animal communication, techniques of study and results of research. *In* "Chemical Systems" (T. A. Sebeok, ed.), pp. 75–102. Univ. of Indiana Press, Bloomington.
710. Wilson, E. O. (1970). Chemical communication within animal species. *In* "Chemical Ecology" (E. Sondheimer and J. B. Simeone, eds.), pp. 133–155. Academic Press, New York.
711. Wilson, E. O. (1971). "The Insect Societies." Harvard Univ. Press, Cambridge, Massachusetts.
712. Wilson, E. O., and Bossert, W. H. (1963). Chemical communication among animals. *Recent Prog. Horm. Res.* **19**, 673–716.
713. Wilson, E. O., and Regnier, F. E., Jr. (1971). The evolution of the alarm-defense system in the formicine ants. *Am. Nat.* **105**, 279–289.
714. Wilson, J., Kuehn, R., and Beach, F. (1963). Modification of the sexual behavior of male rats produced by changing the stimulus female. *J. Comp. Physiol. Psychol.* **56**, 636–644.
715. Wood, D. L. (1962). The attraction created by males of a bark beetle *Ips confusus* (LeConte) attacking ponderosa pine. *Pan-Pac. Entomol.* **38**, 141–145.
716. Wood, D. L. (1970). Pheromones of bark beetles. *In* "Control of Insect Behavior by Natural Products" (D. L. Wood, R. M. Silverstein, and M. Nakajima, eds.), pp. 301–316. Academic Press, New York.
717. Wood, D. L. (1972). Selection and colonization of ponderosa pine by bark beetles. *Symp. R. Entomol. Soc. London* **6**, 101–117.
718. Wood, D. L., and Bushing, R. W. (1963). The olfactory response of *Ips confusus* (LeConte) (Coleoptera:Scolytidae) to the secondary attraction in the laboratory. *Can. Entomol.* **95**, 1066–1078.
719. Wood, D. L., Browne, L. E., Silverstein, R. M., and Rodin, J. O. (1966). Sex pheromones of bark beetles. I. Mass production, bio-assay, source, and isolation of the sex pheromone of *Ips confusus* (LeC.) *J. Insect Physiol.* **12**, 523–536.
720. Wrede, W. L. (1932. Versuche über den Artduft der Elritzen. *Z. Vergl. Physiol.* **17**, 510–519.
721. Wright, R. H. (1958). The olfactory guidance of flying insects. *Can. Entomol.* **90**, 81–89.
722. Wright, R. H. (1962). The attraction and repulsion of mosquitos. *World Rev. Pest Control* **1**, 2–12.
723. Wright, R. H. (1964). "The Science of Smell." Allen & Unwin, London.
724. Wynne-Edwards, V. C. (1962). "Animal Dispersion in Relation to Social Behaviour." Oliver & Boyd, Edinburgh and London.
725. Young, J. C., Silverstein, R. M., and Birch, M. C. (1973). Aggregation pheromone of the beetle *Ips confusus:* Isolation and identification. *J. Insect Physiol.* **19**, 2273–2277.
726. Zenther-Moller, O., and Rudinsky, J. A. (1967). Studies on the site of sex pheromone production in *Dendroctonus pseudotsugae* (Coleoptera:Scolytidae). *Ann. Entomol. Soc. Am.* **60**, 575–582.

Taxonomic Index

Taxonomic information is included in parentheses after each indexed animal name.

VERTEBRATES

BIRDS, 2, 42
    Leach's petrel (*Oceanodroma leucor-rhoa*), 42
FISH, 17, 42, 55, 61, 68, 70, 81, 94
    Blind goby (*Typhlogobius californien-sis*), 38
    Char (*Salmo alpinus*), 42
    Cypriniform fish (Cyprinidae), 69
    Goby (*Bathygobius soporator*), 82, 94
    Guppy (*Poecilia reticulata*), 118
    Minnows (Cyprinidae), 68, 69
    Salmon (*Salmo*), 42
    Yellow bullhead (*Ictalurus natalis*), 38, 73, 81
MAMMALS
Artiodactyla
    Bison (*Bison bonarus*), 13
    Black-tailed deer (*Odocoileus hemionus columbianus*), 37, 73
    Musk deer (*Mochus moschiferus*), 102
    Pronghorn antelope (*Antilocapra americana*), 14
    Ruminants, 11
    Sheep, ram (*Ovis aries*), 35
    Swine (*Sus*), 94, 102, 118, 119
Carnivora
    Brown bears (*Ursus*), 13, 14
    Canidae, 13, 43
    Civet cat (*Viverra civetta*), 102
    Dog (*Canis*), 13, 17, 21, 38, 50
    Hyena (*Crocuta crocuta*), 13

    Mongoose (*Herpestes*), 12
    Wolf (*Canis*), 13
Insectivora
    Tree shrew (*Tupaia bellangeri*), 40
Lagomorpha
    Rabbit, 1, 12, 13, 38, 40, 84
       Oryctolagus cuniculus, 36, 68
Marsupialia
    Gliding phalanger (*Petaurus breviceps papuanus*), 40
Perissodactyla
    Black rhinoceros (*Diceros bicornis*), 44
Primates, 2, 101
    Apes (Pongidae), 101
    Lemur (*Lemur catta*), 83
    Loris (*Nycticebus coucang*), 44
    Man (*Homo sapiens*), 1, 101, 102, 118, 119
    Monkeys (Anthropoidae), 101
      Marmoset
        Callithrix j. jacchus, 8
        Leontopithecus rosalia rosalia, 12
      Rhesus (*Macaca*), 94, 97, 103
      Squirrel (*Saimiri*), 108
Rodentia, 37
    Beaver (*Castor canadensis*), 12, 36
    Guinea pig (*Cavia porcellus*), 73, 83
    Mongolian gerbil (*Meriones unguiculatus*), 13
    Mouse (*Mus*), 4, 38, 39, 54, 67, 68, 73, 81, 82, 111, 120

159

Rat (*Rattus*), 35, 54, 73
Syrian golden hamster (*Mesocricetus auratus*), 50, 54
REPTILES
Snakes, 21, 50, 51, 60

Western banded gecko (*Coleonyx variegatus*), 94
Tortoise, 97
Geochelone, 94, 98

INVERTEBRATES

ACRASIALES
Cellular slime mold, 60, 65, 117
   Dictyostelium discoideum, 60, 61, 66
ANNELIDA
Oligochaeta
   Earthworm (*Lumbricus terrestris*), 69
Polychaeta
   Polychaete worm (*Platynereis dumerilii*), 97
ARTHROPODA (*See also* INSECTA), 16, 20, 50, 51, 108
Acarina
   Mites, 51, 93
   Ticks, 63
Arachnida
   Spiders, 51
   lycosid (Lycosidae), 94
Crustacea, 109
   Amphipods, 46
   Barnacles (*Balanus*), 55
   Crabs, 93, 118
   Cancer, 112
   hermit (*Pagurus bernhardus*), 94
   Pachygrapus, 112
   Desert wood louse (*Hemilepistus reaumuri*), 38
   Terrestrial isopods (wood lice), 60
Diplopoda
   Millipedes (Arthropoda:Diplopoda), 95
ASCHELMINTHES
Nematoda
   Nematodes, 15, 50, 51, 97
Rotifera
   Rotifers (*Brachionus*), 50, 97, 98
ECHINODERMATA
Echinoidea
   Sea urchin (*Diadema antillarium*), 69, 70
INSECTA
Coleoptera (beetles), 19, 62, 86
   Ambrosia beetles (Scolytidae), 56

Anthicid beetles (Anthicidae), 62
Bark beetles (Scolytidae), 26, 56, 58, 62, 73, 74, 82, 119
Black carpet beetle (*Attagenus megatoma*), 86
Blaps sulcata, 69
Boll weevil (*Anthonomus grandis*), 62
California 5-spined ips (*Ips confusus*), 58
Coccinellid beetles (Coccinellidae), 60
Douglas fir beetle (*Dendroctonus pseudotsugae*), 58, 74
Flour beetle (*Tribolium confusum*), 66
Grass grub beetle (*Costelytra zealandica*), 86
Ground beetle (*Pterostichus lucublandus*), 73
Ips, 119
Japanese beetle (*Popillia japonica*), 62
Lycus loripes, 61
Meloid beetles (Meloidae), 62
Red flour beetle (*Tribolium castaneum*), 66
Southern pine beetle (*Dendroctonus frontalis*), 58, 74
Sugar-beet wireworm (*Limonius californicus*), 86
Trogoderma inclusum, 86
Weevil, *Pissodes*, 62
Western pine beetle (*Dendroctonus brevicomis*), 58
Yellow mealworm beetle (*Tenebrio molitor*), 72
Collembola (springtails), 98
Sinella curviseta, 98
Diptera (flies), 93
   Apple maggot fly (*Rhagoletis pomonella*), 72
   Crab hole mosquito (*Deinocerites cancer*), 51
   Culex tarsalis, 55

Drosophila, 111
Drosophila melanogaster, 23, 119
Housefly (*Musca domestica*) , 86, 98, 99
Mediterranean fruit fly (*Ceratitis capitata*) , 93
Mosquitoes, 23, 55, 93
Queensland fruit fly (*Dacus tryoni*) , 95
Sheep blowfly (*Lucilia cuprina*) , 55, 119
Hemiptera
Bedbug (*Cimex lectularius*) , 60, 69, 70
Dysdercus intermedius, 69
Nabid bugs (Nabidae) , 71
Homoptera
Aphid, 69
Acyrthosiphon pisum, 71
Hymenoptera (ants, bees, wasps) , 19–22, 25, 28, 35, 41, 45, 46, 52, 53, 68–71, 75–77, 83, 86
Acanthomyops, 76
Army ant (*Leptogenys*) , 46
Atta cephalotes, 47
Bumble bee (*Bombus*) , 51
Camponotus, 76
Camponotus herculeanus, 85
Camponotus ligniperda, 37
Crematogaster, 76
Dolichoderinae, 46, 76
Dolichoderus, 76
Dorylinae, 46
Formica, 76, 80, 83
Formicinae, 46, 76
Harvester ant (*Pogonomyrmex badius*) , 46
Honeybee (*Apis mellifera*) , 1, 4, 43, 48, 53, 54, 80, 82, 86
Imported fire ant (*Solenopsis saevissima*) , 21, 29
Iridomyrmex, 76
Manica, 76
Messor, 76
Mycocepurus, 76
Myrmica, 76
Myrmica rubra, 83
Myrmicinae, 46, 76
Odontomachus, 76
Paltothyreus, 76
Phaeogenes cynarae, 72
Pharaohs ant (*Monomorium pharaonis*) , 47
Pogonomyrmex, 76
Pogonomyrmex badius, 28

Ponerinae, 46, 76
Solitary bee (*Augochiora pura*) , 63
Stingless bee
Lestrimelitta, Melipona, Trigona, 47
Lestrimelitta limao, 47
Trigona postica, 48
Tapinoma, 76
Telenomus sphingis, 72
Texas leaf-cutting ant (*Atta texana*) , 29, 30, 47
Trichogramma evanescens, 72
Wasp (*Vespa vulgaris*) , 43
Isoptera (termites) , 48, 51, 52, 75, 80
Drepanotermes rubriceps, 80
Kalotermes flavicollis, 53
Nasutitermes, 48, 80
Nasutitermes exitiosus, 49, 79
Zootermopsis nevadensis, 48
Lepidoptera (butterflies, moths) , 9, 17, 19, 23, 24, 26, 27, 33, 66, 85, 92, 95–97, 106, 107, 109, 113–115
Aegeriidae 86
Alfalfa looper moth (*Autographa californica*) , 113, 114, 116
Almond moth (*Cadra cautella*) , 87
Angoumois grain moth (*Sitotroga cerealella*) , 86
Arctiid moth (*Holomelina immaculata*) , 106
Arctiidae, 86
Beet armyworm (*Spodoptera exigua*) , 86
Bombycidae, 86
Cabbage looper moth (*Trichoplusia ni*) , 5, 9, 15, 25–27, 31, 32, 86, 98–100, 106, 107, 110, 113, 114, 116
Clepsis spectrana, 87
Codling moth (*Laspeyresia pomonella*) , 87
Cotton bollworm (*Heliothis zea*) , 86
Creatonetus gangis, 15
Eastern spruce budworm (*Choristoneura fumiferana*) , 87
Eucosmidae, 86
European corn borer (*Ostrinia nubilalis*) , 87, 115
European pine shoot moth (*Rhyacionia buoliana*) , 87
Fall armyworm (*Spodoptera frugiperda*) , 86
False codling moth (*Argyroploce leucotreta*) , 86

Lepidoptera (cont.)
Fruit-tree leaf roller (Archips argyrospilus), 87, 115
Fruit-tree tortrix (Archips podana), 87, 115
Gelechiidae, 86
Grape berry moth (Paralobesia viteana), 87
Greater wax moth (Galleria mellonella), 93
Gypsy moth (Porthetria dispar), 31, 86
Indian-meal moth (Plodia interpunctella), 87
Lesser peach tree borer (Synanthedon pictipes), 86, 115
Lesser wax moth (Achroia grisella), 93
Light brown apple moth (Epiphyas postvittana), 101, 106
Lymantriidae, 86
Mediterranean flour moth (Anagasta kuehniella), 24, 87
Noctuidae, 86
Oblique-banded leaf roller (Choristoneura rosaceana), 87
Omnivorous leaf roller (Platynota stultana), 115
Oriental fruit moth (Grapholitha molesta), 87
Peach tree borer moth (Sanninoides exitiosa), 86, 115
Pine emperor moth (Nudaurelia cytherea cytherea), 87
Pink bollworm moth (Pectinophora gossypiella), 25–27, 86, 108
Pyralidae, 87
Queen butterfly (Danaus gilippus berenice), 96
Raisin moth (Cadra figulilella), 87
Red-banded leaf roller (Argyrotaenia velutinana), 87, 115
Red bollworm (Diparopsis castanea), 87
Saturniidae, 87, 112
Silkworm moth (Bombyx mori), 4, 17, 86

Smaller tea tortrix (Adoxophyes fasciata), 87
Smartweed borer (Ostrinia obrumbratalis), 115
Southern armyworm moth (Spodoptera eridania), 87, 97
Soybean looper (Pseudoplusia includens), 86
Spodoptera littoralis, 87
Spodoptera litura, 87
Summer fruit tortrix (Adoxophyes orana), 87
Tiger moths (Arctiidae), 86
Tobacco budworm (Heliothis virescens), 86
Tobacco moth (Ephestia elutella), 87
Tortricidae, 87
Tufted apple bud moth (Platynota idaeusalis), 87
Orthoptera
Blaberus craniifer, 62
Cockroaches (Blattidae), 62, 95, 111
Desert locust (Schistocerca gregaria), 55
Migratory locust (Locusta migratoria migratorioides), 62
MOLLUSCA
Bivalvia
Oyster (Pinctada), 55
Gastropoda
Limpets, 42
Snails, 21, 69
Crepidula, 51
Fasciolaria tulipa, 49, 66, 67
Littorina littorea, 63
pulmonate, 22
PLATYHELMINTHES
Turbellaria
Planaria, 63, 109
PROTOZOA, 2, 50
Ciliata
Ciliate protozoan (Rhabdostyla vernalis), 50

Subject Index

A

Acetic acid, 97, 103
Active space, 28–30
3,5–Adenosine monophosphate, 60
Aerial odor trails, 22–29, 42, 47, 48, 50, 51
Age and pheromone behavior, 11, 38, 108
Aggregation pheromones, 5, 19, 26, 36, 45, 74, 82, 85
Aggression pheromones, 5, 8, 11, 14, 45, 52, 75–77, 79, 80, 83, 118
Aggression toward a conspecific, 80
Air velocity, 105–107
Alarm pheromone behavior, 6, 11, 28, 52, 54, 69, 70, 75–77, 79, 80, 83, 118
o-Aminoacetophone, 76
Androgen, 39, 111, 118, 119
5-Androst-16-en-3-one, 95
Anemotaxis, 23–27, 33
Antiaggregation, 57, 71
Antiaphrodisiacs, 72
Aphrodisiacs, 93
Appraisals of pheromones, 35
Arrestment of locomotion, 19, 27, 45, 50, 65, 93, 95–97
Assumption of copulatory position, 4

B

Bacteria, synthesis of pheromone by, 3
Benzaldehyde, 97
Biosynthetic mechanisms, 3
Breeding conditions affecting pheromone behavior, 38
endo-Brevicomin, 58, 74

exo-Brevicomin, 58
Butanoic acid, 97, 103

C

n-Caproic acid, 48
Castoreum glands, 36
Chemoklinotaxis, 20, 24
Chemotactic behavior, 24–26
Chemotactic trail following, 33
Chemotropotaxis, 20, 24
Chin glands, 12, 13, 36
Circadian rhythms, 108, 109, 112
Circumgenital glands, 12
Citral, 47
Citronellal, 76
Civetone, cis-9-cycloheptadecenone, 103
Colonies, 35, 41, 43, 54
Communication, 1
Communication distances, 28–31
Concentration gradients, 19–21, 23
Control of pests, 4, 5
Copulatory behavior, 26, 50, 85, 97, 101, 110
Copulatory readiness, 38
Coremata, 10
Corpora allata, 111
Courtship behavior, 50, 85, 93, 94, 96, 98, 101, 113, 114
Crustecdysone, 93, 112, 118

D

n-Decane, 76
3-Decanone, 76

163

cis-5-Decenyl isovalerate, 87
Diffusion, 15
2,3-Dihydro-7-methyl-*H*-pyrrolizin-1-one, 96
2,6-Dimethyl-3-*n*-butylpyrazine, 76
cis-3,3-Dimethyl-△¹α-cyclohexaneacetaldehyde, 63
trans-3,3-Dimethyl-△¹α-cyclohexaneacetaldehyde, 63
cis-3,3-Dimethyl-△¹β-cyclohexaneethanol, 63
Dimethyl disulfide, 76
2,5-Dimethyl-3-isopentylpyrazine, 76
4,6-Dimethyl-4-octen-3-one, 76
2,6-Dimethyl-3-*n*-pentylpyrazine, 76
Dimethyl trisulfide, 76
trans, trans-3,7-Dimethyldeca-2,6-dien-1,10-diol, 96
Dispersion-inducing pheromones, 5, 36, 52, 68, 69, 120
Distances of pheromone communication, 27, 32, 33
Dominance hierarchies, 11, 38–40, 68, 73, 80, 81
trans-8,trans-10-Dodecadienol, 87
trans-9,11-Dodecadienyl acetate, 87
n-Dodecane, 76
cis-7-Dodecenyl acetate, 86, 116
cis-8-Dodecenyl acetate, 87
cis-9-Dodecenyl acetate, 87
trans-7-Dodecenyl acetate, 86
trans-9-Dodecenyl acetate, 87
11-Dodecenyl acetate, 87
Dodecyl acetate, 87
Dufour's gland, 36, 46, 47, 80

E

Ecology, 5
Ectohormone, 2
Emission of pheromones, 7
Environmental control of pheromone behavior, 93, 105
Epidermal glands, 118
cis-7,8-Epoxy-2-methyloctadecane, 86
Estrogen, 111, 118
Evolution of pheromonal communication, 117
Exaltolide, 15-pentadecanolide, 102
Exploratory trails, 46

F

Family groups, 52
trans-β-Farnesene, 69
Feces and pheromone marking, 11–13, 84
Feeding attractants, 2
Flight speed, 32
Footprint substance, 43
Foraging behavior, 45
Formic acid, 76
Frontalin, 58

G

Geographic areas and reproductive isolation, 113
Geranial, 47, 76
Geranic acid, 48
Geraniol, 48
Gustation, 7, 16, 22

H

Habituation to pheromones, 110, 111
2-Heptanone, 76
Hexanal, 70
trans-10,cis-12-Hexadecadienol, 86
cis-7,cis-11-Hexadecadienyl acetate, 86
cis-7,trans-11-Hexadecadienyl acetate, 86
cis-11-Hexadecenal, 86
2-Hexenal, 76
trans-2-Hexenal, 70
Hibernation, 60
Hierarchies of sex pheromone behavior, 98, 101
Hive atmosphere, 43
Home range, 35, 38, 41, 43, 44
Hormonal steroids, 119
Hormones and pheromone behavior, 4, 39, 108, 111, 112, 118
Human sex pheromones, 101
3α-Hydroxy-5α-androst-16-ene, 95
9-Hydroxydec-trans-2-enoic acid, 54
cis-4-Hydroxydec-6-enoic acid lactone, 37

I

Identification of colony, 43
Identification of individual animals, 11, 14, 35, 36, 38

Identification of physiological status, 35, 38, 39
Immobilization reflex, 94
Imprinting, 37
Inguinal glands, 36
Inhibition of aggregation, 82
Inhibition of aggression, 57
Ipsdienol, 58, 119
Ipsenol, 58, 119
(+) -cis-2-Isopropenyl-1-methylcyclobu-taneethanol, 63
12-Isopropyl-1,5,9-trimethylcyclotetradeca-1,5,9-triene, 49
Isovaleric acid, 13

L

Labia, 102
Light intensity and pheromone behavior, 105, 106
Limonene, 80
Locomotion rate, 5, 33

M

Maintenance of spacing between animals, 65
Maintenance of territories, 67
Male attendance behavior, 93, 94
Mandibular glands, 47, 48, 52, 53, 66, 85
Maternal pheromone, 35, 54
Metathoracic tibial glands, 46
2-Methoxy-5-ethylphenol, 62
5-Methyl-3-butyl octahydroindolizine, 47
3-Methyl-2-cyclohexenol, 58
3-Methyl-cyclopentadecanone, 102
3-Methyl-2-cyclohexenone, 58, 74
2-Methylheptadecane, 86
2-Methyl-4-heptanone, 76
4-Methyl-3-heptanone, 76
6-Methyl-5-hepten-2-one, 76
(—) -14-Methyl-cis-8-hexadecenol, 86
4-Methyl-2-hexanone, 76
4-Methyl-3-hexanone, 76
(—) -Methyl-14-methyl-cis-8-hexadeceno-ate, 86
Methyl-4-methylpyrrole-2-carboxylate, 47
Methyl-trans-6-nonenoate, 93
6-Methyl-3-octanone, 76

3-Methylbutanoic acid, 97, 103
4-Methylpentanoic acid, 97, 103
2-Methylpropanoic acid, 97, 103
Morphology, 5
Multiple component pheromones, 92, 115
Muscone, 3-methylcyclopentadecanone, 102
Musk, 102
Myrcene, 59, 119

N

Nasal cavity, 17
Nassanoff gland, 48
Nassanoff pheromone, 54
Neral, 47, 119
Nerolic acid, 48
n-Nonanal, 92
3-Nonanone, 76
trans-6-Nonenoate, 92
trans-6-Nonenol, 92

O

cis-3,cis-13-Octadecadienyl acetate, 86, 115
trans-3,cis-13-Octadecadienyl acetate, 86, 115
n-cis-11-Octadecenal, 92
trans-2-Octenal, 70
3-Octanone, 76, 83
Olfaction, 7, 6, 22
Olfactory thresholds, 29
Orientation to pheromones, 5, 11
Orthokinesis, 19
9-Oxodec-trans-2-enoic acid (queen sub-stance) , 53, 54, 86

P

Parent–young relationships, 36, 45
Pavan's glands, 46
Pedal glands, 11
Perfumes, 102
Phenol, 86
Phenylacetic acid, 13
Pheromone release, 32
Pheromone trails, 26
Photomenotaxis, 22
Physiological clocks, 105, 106

Physiological control of pheromone behavior, 105, 108
α-Pinene, 59, 80
Plant host and pheromone behavior, 105, 107
Poison glands, 46, 47, 80
Polypeptides, 15
Population size indicated by pheromones, 11
Preorbital glands, 12
Prepuce, 102
Preputial glands, 39, 81
Primer pheromones, 3, 51, 53, 62, 120
Prior exposure to pheromones, 108, 110
Prior mating and pheromone behavior, 108, 110
Prolactin, 54
Propanoic acid, 97, 103
Proteins, 15

R

Reception of pheromones, 7, 16
Recognition pheromones, 36
    of group, 39
    of home, 41, 61
    of home range, 43
    of individuals, 37, 120
    of resting site, 41
    of status, 38
Recruitment trails, 46
Release rate of pheromones, 31
Releaser pheromones, 3, 4, 53, 62, 120
Reproductive isolation, 113
Response threshold to pheromones, 31, 33
Retinue behavior, 53
Retrocornal glands, 12
Rheotaxis, 23
Rivalry behavior, 82

S

Salivary glands, 119
Scent brushes, 96, 97
Scent-marking, 9, 11, 13, 14, 29, 36–39, 43, 44, 48, 50, 68, 72, 73, 81, 83
    active, 11
    passive, 11
Seasonal rhythms, 109

Self-identification, 37, 38
Sensory adaptation, 22, 110
Sex attractants, 2, 11
Sex hormones, 81, 83, 111
Sex pheromone behavior, 5, 6, 14, 36, 38, 50, 51, 82, 85, 105, 113, 117, 120
Sexual maturity, 108
Sexual status, 11
Shared pheromones, 116
Skin (cutaneous) glands, 11
Social behavior, 36
Social groups, 35, 40, 52
Social insects, 11, 36, 40, 43, 45, 52, 68, 75, 81, 83
Social status, 35, 36, 38, 39, 73
Sternal glands, 48
Stink fights, 83
Submaxillary glands, 119
Submissiveness, 30, 39, 68, 73, 80, 81
Sun-compass steering, 22
Supraoccipital glands, 12
Swarming behavior, 54
Synchronization of sexual activity, 85

T

Tandem pairing, 50
Tandem running, 47
Tarsal glands, 37, 73
Taxes, 19, 65
Temperature and pheromone behavior, 105, 106, 109, 110
Temporary habitats, 60
Terpene resins, 58, 119
Terrestrial trail pheromones, 20, 21, 28, 29, 35, 42, 45–50, 52, 60, 80
Territoriality, 14, 40, 43, 67, 68, 80–84, 120
Testosterone, 95
trans-3,cis-5-Tetradecadienoic acid, 86
cis-9,trans-11-Tetradecadienyl acetate, 87
cis-9-trans-12-Tetradecadienyl acetate, 86, 87
Tetradecane, 62
cis-9-Tetradecenal, 86
trans-11-Tetradecenal, 87
trans-11-Tetradecenol, 87
11-Tetradecenyl acetate, 115
cis-9-Tetradecenyl acetate, 86, 87
cis-11-Tetradecenyl acetate, 87
trans-11-Tetradecenyl acetate, 87

Transport of pheromone, 14
cis-9-Tricosene, 86
Turbulence, 16

U

n-Undecanal, 92
Undecane, 62
n-Undecane, 76
Urine, 11–14, 38, 44, 50, 68, 73, 81, 82, 112
Urine washing, 44

V

Valeric acid, 86
Ventral glands, 13

cis-Verbenol, 58
trans-Verbenol, 58
Verbenone, 58, 74
Visual orientation, 26

W

Wind velocity, 31–33

X

p-Xylene, 76